# STEVEN CONNORS

# When Science Was Wrong

*A Guide to the Scientific Mistakes of the Past*

*Dedicated to my wife. I hope this book means we can pay rent this month.*

# Contents

# Foreword

In the grand tapestry of human achievement, success and failure are the intertwined threads that weave our collective narrative. This nonfiction exploration delves into the crucible of past failures and emerges with the profound realization that missteps and setbacks are not merely stumbling blocks; they are the very stepping stones that pave the path to triumph. In a world often dazzled by tales of unblemished success, it is imperative to shift our focus to the invaluable lessons encapsulated within our historical blunders. Failure is not a destination but a crucial juncture in the journey toward success.

To illustrate this point, let us venture into the annals of scientific history and recount the tale of one of its most revered figures: Thomas Edison. Widely celebrated for his invention of the electric light bulb, Edison's path to brilliance was not without its share of stumbles. It is often said that he failed a thousand times before successfully illuminating the world. Yet, Edison viewed each failed experiment not as a defeat, but as a discovery of a way that wouldn't work, bringing him one step closer to the ultimate solution. The journey of past failures paving the way for future success is not confined to the realms of invention. It extends to the scientific method itself, a paradigm that embraces the uncertainty of experimentation and the possibility of being wrong.

In the mid-20th century, a towering figure in the scientific community, Albert Einstein, found himself at the center of a heated debate. Einstein's rejection of certain aspects of quantum mechanics, encapsulated in his famous phrase "God does not play dice with the universe," was met with skepticism from other luminaries in the field, most notably Niels Bohr. The clash between Einstein and Bohr symbolized the essence of scientific inquiry—unfettered exploration, the willingness to question even the most established theories, and the humility to accept the prospect of being wrong. In this clash of intellectual titans, the significance of past failures lies not in the preservation of ego but in the advancement of knowledge. Einstein's steadfastness, coupled with the humility to acknowledge the potential fallibility of his beliefs, encapsulates the delicate balance required to transform failures into stepping stones toward progress.

As we navigate the intricate terrain of human history, it becomes evident that the road to success is often lined with the markers of failure. Take, for instance, the story of Steve Jobs, the visionary co-founder of Apple Inc. In 1985, Jobs faced a dramatic ousting from the company he helped create. His departure was not merely a setback but a crushing failure that could have defined the end of his career. However, Jobs' tenacity and ability to learn from his failures fueled his resurgence. A decade later, he returned to Apple, revitalizing the company and reshaping the landscape of personal technology with innovations like the iPod, iPhone, and iPad. The juxtaposition of these stories— from Edison's perseverance in the face of repeated failures to Einstein's openness to reassessing his own theories and Jobs' phoenix-like rise from corporate exile—illuminates a common thread. It is the ability to extract wisdom from the crucible of failure that distinguishes those who ultimately triumph. Failure, when embraced with resilience and a thirst for growth, becomes a formidable catalyst for innovation and achievement.

This exploration of the symbiotic relationship between failure and success is not limited to the realm of individual endeavors. It permeates the fabric of societal progress, where collective missteps serve as cautionary tales and opportunities for collective learning. Consider the financial crises that have punctuated history—the Great Depression of the 1930s, the more recent global financial crisis of 2008. These moments of economic turmoil were undeniably painful, but they also prompted a reevaluation of financial systems and regulatory frameworks, ultimately leading to reforms aimed at preventing similar catastrophes in the future. In the arena of space exploration, the Apollo 1 disaster stands as a stark testament to the human cost of failure. The tragic loss of three astronauts in a pre-flight test was a devastating setback for NASA and the United States. However, the meticulous investigation that followed led to a comprehensive reassessment of safety protocols and engineering practices, resulting in significant improvements that paved the way for subsequent successful lunar missions.

This examination of failure and its transformative potential extends beyond the sciences and economics—it touches the very fabric of human relationships and societal evolution. Nations that confront the darker chapters of their histories, acknowledge past wrongs, and strive for reconciliation exemplify the collective learning that arises from acknowledging failure. The ability to learn from the mistakes of the past is not a sign of weakness but a testament to resilience, adaptability, and the unwavering pursuit of progress. As we embark on this exploration of the interplay between failure and success, it is crucial to dispel the pervasive myth that success is a linear trajectory devoid of setbacks. The stories encapsulated in these pages are not hagiographies of infallible heroes but narratives of resilience, perseverance, and the capacity for growth in the face of adversity. The pages that follow invite you to reconsider failure not as a mark of defeat but as a roadmap to growth, a guidebook for navigating the intricate terrain of ambition and achievement. In the chapters that unfold, we will delve into the psychological nuances of coping with failure, the cultural perspectives that shape our understanding of success and setbacks, and the strategies employed by individuals and organizations to leverage failure as a catalyst for innovation. Through the prism of history, science, and personal anecdotes, we will unravel the intricate dance between past failures and the future successes they herald. In the words of the great Winston Churchill, "Success is not final, failure is not fatal: It is the courage to continue that counts." Let us embark on this journey with the courage to confront our failures, the humility to learn from them, and the unwavering determination to forge a future illuminated by the lessons of the past.

# I

# Medicine

*"The history of medicine is the history of the unusual."*
*-Robert M. Fresco*

# Trepanation

**Understanding Trepanation**

Trepanation, often referred to as trephination or trepanning, is an ancient surgical procedure involving the removal of a portion of the skull, typically by drilling or scraping. This procedure, though rooted in ancient history, is not a scientific theory in the traditional sense but rather a medical practice that has evolved over millennia.

**Origins and Early Development**

Trepanation has a history dating back thousands of years, with evidence of its practice found in various cultures around the world. While the exact origin remains uncertain, it is believed to have been performed as early as the Neolithic period. Some key points about its origins and early development include:

**Neolithic and Prehistoric Practices:** Archaeological findings suggest that trepanation was practiced in regions such as Europe, South America, and Africa during prehistoric times. Skulls bearing evidence of trepanation have been discovered in burial sites, indicating that this procedure was performed for both medical and ritualistic purposes.

**Cultural Variations:** Trepanation was not limited to one culture but was rather a widespread practice. Different cultures had their own techniques and reasons for performing it. Some cultures believed it could treat various ailments, including head injuries, while others thought it had spiritual or supernatural significance.

**Lack of Modern Medical Understanding:** It's important to note that the ancient practice of trepanation lacked the scientific understanding and medical knowledge that we have today. The motivations and beliefs surrounding this practice varied widely, and the effectiveness of trepanation in treating medical conditions was often uncertain.

**Believers and Supporters**

The concept of trepanation has evolved over time, with different cultures and historical periods having various beliefs and motivations for performing this procedure. Believers and practitioners of trepanation included:

**Ancient Cultures:** Various ancient cultures, including the Egyptians, Greeks, and Incas, practiced trepanation. They believed it could alleviate headaches, seizures, and other neurological conditions, although the scientific basis for these beliefs was lacking.

**Shamans and Healers:** In many societies, shamans, tribal healers, or medicine men were responsible for performing trepanation. It was often seen as a ritualistic practice with spiritual significance.

**The Story of How the Theory Eventually Got Disproved**

Trepanation's decline as a medical practice can be attributed to several factors, including advancements in medical knowledge and a better understanding of the human body:

**Emergence of Modern Medicine:** The practice of trepanation began to decline with the emergence of modern medicine and the development of more effective treatments for head injuries and neurological conditions. As medical knowledge expanded, trepanation was no longer viewed as a viable medical solution.

**Scientific Understanding of Brain Function:** As science advanced, there was a growing understanding of the brain's structure and function. Medical professionals began to recognize that trepanation, as it was practiced in the past, lacked a scientific basis and often posed significant risks to patients.

**Improved Surgical Techniques:** Modern surgical techniques and tools made trepanation unnecessary for most medical conditions. The risks associated with the procedure, including infection and damage to vital brain structures, outweighed any potential benefits.

**When Trepanation Got Disproved**

The disapproval of trepanation as a medical practice occurred gradually over centuries:

**Ancient to Medieval Times:** Trepanation was performed in various forms across cultures for millennia, but its decline as a mainstream medical practice began in antiquity and continued into the Middle Ages.

**Early Modern Period:** By the early modern period, trepanation had largely fallen out of favor in Europe as advancements in medical science and surgery provided more effective treatments for head injuries and neurological conditions.

**Contemporary Medicine:** In contemporary medicine, trepanation is considered an obsolete and risky procedure. It is no longer practiced as a legitimate medical treatment for neurological disorders or head injuries.

While trepanation has a long and varied history, it is

important to emphasize that modern medicine has moved far beyond this ancient practice. Trepanation, once performed with ritualistic and often unscientific motivations, is now viewed as a historical curiosity rather than a viable medical treatment. Advances in neuroscience, surgery, and medical knowledge have provided us with safer and more effective approaches to treating neurological conditions and head injuries, rendering trepanation obsolete in the modern medical landscape.

## Reflexology: An Examination of a Disproved Alternative Therapy

Reflexology, a practice rooted in the belief that pressure applied to specific points on the feet, hands, or ears can affect various organs and systems of the body, has been a topic of interest and debate within the realm of alternative medicine. This theory suggests that these "reflex points" correspond to different parts of the body and that manipulating them can promote healing and well-being. In this exploration, we will delve into the origins of reflexology, its proponents, its rise to popularity, the scientific skepticism it faced, and its eventual debunking as an evidence-based therapy.

**Origins of the Theory:** Reflexology, often associated with traditional Chinese medicine and ancient Egyptian practices, claims to be rooted in ancient healing traditions. The fundamental premise is that the body is mapped onto the feet, hands, and ears, with each area corresponding to specific organs, glands, or body systems. By applying pressure to these reflex points, practitioners believe they can stimulate energy flow, promote healing, and alleviate various health issues.

**Modern Revival and Eunice Ingham:** The modern revival of reflexology can be largely attributed to Eunice Ingham, an American nurse and physiotherapist who practiced in the mid-20th century. Ingham is often credited with mapping out the reflex points on the feet and developing the foot reflexology techniques that are commonly used today.

In the 1930s, Ingham published her work, "Stories the Feet Can Tell," which outlined her theories and practical applications of reflexology. She believed that by manipulating the reflex points on the feet, one could detect and treat health issues in corresponding areas of the body.

**Belief in Reflexology:** Reflexology gained popularity, particularly within the realm of alternative and complementary medicine. Advocates and practitioners claimed that this therapy could alleviate a wide range of ailments, from headaches and digestive problems to stress and insomnia. Reflexology charts, illustrating the reflex points on the feet, hands, and ears, became common tools for practitioners.

**Scientific Skepticism:** Reflexology faced skepticism from the scientific and medical communities. The primary criticisms included:

**Lack of Scientific Basis:** Reflexology lacked a scientifically established mechanism of action. Critics argued that the concept of reflex points and energy flow was not supported by empirical evidence or physiological principles.

**Inconsistent Claims:** Practitioners often made sweeping claims about reflexology's effectiveness in treating a wide range of health issues, but these claims were not substantiated by rigorous clinical trials or controlled studies.

**Placebo Effect:** Some of the benefits attributed to reflexology were seen as potentially related to the placebo effect, where individuals experienced improvements due to the belief in the therapy rather than its physiological effects.

**The Discrediting of Reflexology:** The gradual discrediting of reflexology as an evidence-based therapy was a result of several factors:

**Lack of Scientific Support:** Despite decades of practice and anecdotal claims, reflexology failed to produce compelling scientific evidence to support its underlying principles and therapeutic efficacy. Research studies consistently lacked robust methodology and failed to demonstrate consistent, replicable results.

**Contradictory Claims:** Practitioners often made contradictory claims about the precise reflex points and their corresponding organs or systems. The lack of standardization and consensus within the field contributed to skepticism.

**Mainstream Medical Rejection:** Reflexology did not gain acceptance within mainstream medicine. Medical professionals remained unconvinced of its efficacy, and it was not integrated into standard medical practice.

**Shift Towards Evidence-Based Medicine:** As the medical field increasingly emphasized evidence-based practices, therapies like reflexology, lacking scientific substantiation, faced declining acceptance.

**Modern Status of Reflexology:** Today, reflexology exists primarily within the domain of alternative and complementary medicine. While it still has a dedicated following and some individuals report experiencing relaxation and stress relief from reflexology sessions, it is not considered a substitute for evidence-based medical treatments.

The scientific and medical communities continue to view reflexology with skepticism, primarily due to the lack of empirical support for its fundamental principles. Although some studies have explored reflexology's potential benefits in alleviating certain symptoms or enhancing relaxation, the overall body of evidence remains inconclusive.

**Conclusion:** Reflexology, a therapy based on the premise that specific points on the feet, hands, or ears correspond to organs and systems in the body, had its origins in ancient healing traditions and experienced a modern revival through the work of Eunice Ingham. Despite gaining popularity in the realm of alternative medicine, it faced skepticism from the scientific and medical communities due to a lack of empirical evidence and inconsistent claims.

Reflexology's decline as an evidence-based therapy can be attributed to the absence of scientific support, contradictory claims, and the shift towards evidence-based medicine in the medical field. While it continues to have a following and is appreciated by some for its relaxation benefits, it is not considered a substitute for proven medical treatments. Reflexology serves as a reminder of the importance of rigorous scientific scrutiny and evidence-based practices in the evaluation of alternative therapies.

*Miasmatic Theory of Disease: Unraveling the Belief in Noxious Air*

The miasmatic theory of disease was a prevailing medical hypothesis that attributed the spread of diseases, particularly epidemic and infectious diseases, to foul-smelling, noxious air known as "miasma." Originating in ancient times, this theory endured for centuries and was embraced by prominent figures in the medical field. It wasn't until the mid-19th century that the theory was discredited, giving way to a more accurate understanding of the role of microorganisms in disease transmission. In this exploration, we will delve into the essence of the miasmatic theory, its historical origins, its proponents, the influential individuals who believed in its validity, the process by which it was ultimately debunked, and the pivotal moment when it was conclusively disproved.

## The Essence of Miasmatic Theory

The miasmatic theory of disease posited that diseases, particularly contagious and epidemic illnesses, were caused by foul-smelling, noxious air or "miasma." It was believed that these noxious vapors arose from decomposing organic matter, polluted water, or other environmental sources. Exposure to miasma, according to the theory, was thought to be the primary means by which individuals contracted diseases.

## Origins and Early Notions of Miasma

The concept of miasma dates back to ancient civilizations, where it was linked to beliefs about the malevolent influence of noxious air. In ancient Greece and Rome, for example, miasma was associated with "bad air" and was thought to be a cause of various diseases.

**Hippocrates and Humoral Theory:** The influential

Greek physician Hippocrates, often regarded as the father of medicine, contributed to early notions of miasma. His humoral theory of medicine, which dominated medical thought for centuries, incorporated the idea that imbalanced bodily fluids could make individuals susceptible to miasma and disease.

**Proponents of Miasmatic Theory**

The miasmatic theory gained traction and was endorsed by influential figures in the field of medicine and public health. These proponents believed that miasma was a key factor in the spread of diseases.

**John Snow's Cholera Studies:** In the 19th century, English physician John Snow conducted pioneering research on the transmission of cholera. While he is now celebrated for his work in identifying contaminated water sources as the cause of cholera outbreaks, his initial investigations were influenced by miasmatic theory, as he believed in the role of foul air in disease transmission.

**The Decline of Miasmatic Theory: Emergence of Germ Theory**

The miasmatic theory began to face scientific challenges as new discoveries in the fields of microbiology and epidemiology emerged. The breakthrough came with the acceptance of germ theory, which proposed that microorganisms, such as bacteria and viruses, were the actual agents responsible for causing diseases.

**Germ Theory's Emergence:** The mid-19th century saw the rise of germ theory, championed by scientists like Louis Pasteur and Robert Koch. This theory introduced a groundbreaking paradigm shift, asserting that microorganisms were the causative agents of infectious diseases.

**Louis Pasteur's Experiments:** Louis Pasteur's experiments, including his work on fermentation and the pasteurization process, provided compelling evidence for the role of microorganisms in disease. His discoveries led to the recognition of the connection between specific bacteria and illnesses.

**Robert Koch's Postulates:** Robert Koch, a German physician, developed a set of postulates that linked specific microorganisms to particular diseases. His work on tuberculosis and cholera helped solidify the principles of germ theory.

### The Disproof of Miasmatic Theory

As germ theory gained acceptance, miasmatic theory began to wane in credibility. The empirical evidence and experimental methods associated with the new theory provided a more accurate and testable framework for understanding disease transmission.

**Cholera Outbreak Investigations:** The cholera outbreaks of the mid-19th century offered a stark contrast between the explanations of miasmatic theory and germ theory. John Snow's famous investigations in London, where he traced cholera cases to contaminated water sources, highlighted the limitations of miasmatic theory in explaining the transmission of infectious diseases.

**Advances in Microbiology:** The development of microscopy and the ability to visualize microorganisms played a crucial role in discrediting miasmatic theory. Researchers could now directly observe the presence of bacteria and viruses in diseased tissues.

### The Conclusive Discrediting of Miasmatic Theory

By the late 19th century, miasmatic theory had been conclusively discredited in favor of germ theory. The understanding that microorganisms were the agents responsible for diseases revolutionized medicine and led to significant advancements in public health, sanitation, and the development of vaccines and antibiotics.

## Legacy and Insights

The decline of miasmatic theory and the rise of germ theory represented a transformative shift in the field of medicine. It highlighted the importance of empirical evidence and the scientific method in understanding the causes of diseases. The discrediting of miasmatic theory paved the way for modern epidemiology and the development of effective strategies for disease prevention and treatment, ultimately saving countless lives.

# Tapeworm Diet

The theory of the tapeworm diet is a bizarre and potentially dangerous concept that emerged in the late 19th and early 20th centuries. It is based on the idea that intentionally ingesting a tapeworm could lead to weight loss, as the tapeworm would consume some of the food ingested by the host, thereby reducing the host's calorie intake. This misguided theory gained popularity during an era when extreme and unproven weight-loss methods were prevalent.

**Origins of the Theory**

The origins of the tapeworm diet theory can be traced back to the late 19th century when society's obsession with thinness and physical appearance began to intensify. In this era, beauty standards for women, in particular, placed a high premium on a slim figure, leading to various efforts to achieve this ideal.

While it is challenging to pinpoint a single individual who first proposed the tapeworm diet, the idea likely emerged as part of the broader culture of unproven and extreme weight-loss methods that were prevalent during this time. The concept may have been inspired by anecdotal reports of accidental tapeworm infestations leading to weight loss.

**Prominent Believers in the Theory**

The theory of the tapeworm diet did attract some believers,

although it is essential to note that these individuals were not reputable scientists or medical professionals. Instead, they were often individuals or entities with vested interests in promoting dubious weight-loss products and methods. This included certain advertisements and pamphlets that peddled tapeworm capsules as a solution for weight loss.

However, the theory was widely discredited and condemned by mainstream medical practitioners and scientists. Prominent figures in the medical community denounced the tapeworm diet as both scientifically unsound and potentially hazardous to one's health.

## The Disproof of the Theory

The theory of the tapeworm diet began to unravel as medical knowledge advanced, and the risks associated with tapeworm infestations became more apparent. There were several key factors that contributed to the disproof of this dieting method.

**Health Risks**: As scientific understanding of parasitic infections improved, it became increasingly evident that tapeworm infestations could lead to severe health complications. Tapeworms can cause malnutrition, digestive problems, and even life-threatening conditions if left untreated. This knowledge raised alarm about the potential dangers of intentionally ingesting tapeworms.

**Lack of Scientific Evidence**: The tapeworm diet was never supported by credible scientific evidence. While proponents touted anecdotal weight loss stories, there was no rigorous scientific research to substantiate these claims. The absence of empirical data and the potential health risks associated with tapeworm infestations undermined the diet's credibility.

**Medical Warnings**: The medical community widely condemned the tapeworm diet, with physicians and public health officials issuing warnings against its use. These warnings emphasized the dangers posed by tapeworm infestations and discouraged individuals from attempting this unproven and risky weight-loss method.

**Legislation and Regulation**: In response to the proliferation of tapeworm capsules and other questionable weight-loss products, governments began to enact legislation to regulate their sale and distribution. This regulatory action aimed to protect the public from potentially harmful products.

## The Disproof of the Theory in Modern Times

The tapeworm diet is now widely recognized as a dangerous and unethical practice. In modern times, it is not only discredited but also illegal in most countries. The awareness of the health risks associated with tapeworm infestations, coupled with strict regulations on the sale and distribution of tapeworms, has effectively put an end to this hazardous weight-loss theory.

Today, tapeworm infestations are treated as serious medical conditions that require prompt medical intervention. Public health agencies and medical professionals strongly advise against attempting to use tapeworms or any other parasites for weight loss.

In summary, the theory of the tapeworm diet, which advocated intentional ingestion of tapeworms as a weight-loss method, emerged during a period of extreme beauty standards and unproven weight-loss methods in the late 19th and early 20th centuries. While some individuals and entities promoted this theory, it was widely discredited and condemned by the mainstream medical community due to its lack of scientific evidence and potential health risks. The disproof of the theory was driven by a growing understanding of the dangers of tapeworm infestations, lack of scientific support, medical warnings, and legislative action to regulate weight-loss products. In modern times, the tapeworm diet is not only discredited but also illegal in most countries, highlighting the recognition of its inherent dangers.

# Cranial Vibrations

The theory of cranial vibrations, a concept rooted in the belief that subtle rhythmic movements of the cranial bones play a vital role in maintaining health and well-being, provides a unique perspective on the historical evolution of medical theories and practices. This theory posits that these cranial movements are linked to the flow of vital energy or life force throughout the body, echoing ancient principles from traditional healing systems. Despite its unconventional nature, the theory of cranial vibrations gained traction in the early 20th century and attracted attention from notable figures within the osteopathic community. However, as the medical field advanced and evidence-based practices gained prominence, the theory gradually lost credibility and was ultimately disproven.

**Origins of the Theory**

The theory of cranial vibrations emerged during a period of significant exploration and experimentation within the field of medicine in the late 19th and early 20th centuries. It was closely associated with the development of cranial osteopathy, a branch of osteopathic medicine that focused on the cranial bones and their purported movements. The origins of this theory are challenging to attribute to a single individual, as it likely evolved as part of a broader culture of unconventional medical ideas during this era.

While it's challenging to pinpoint a single originator of the theory, William Garner Sutherland, an American osteopath, played a significant role in its development. Sutherland, who lived from 1873 to 1954, is often credited with pioneering cranial osteopathy and elaborating on the concept of cranial rhythms and movements. He reported that his inspiration came from a moment of insight during a skull examination, where he claimed to have observed subtle cranial movements that had previously gone unnoticed.

**Prominent Believers in the Theory**

The theory of cranial vibrations did attract some proponents, particularly within the osteopathic community and among practitioners of alternative and holistic medicine. Individuals like Sutherland and his contemporaries believed that these subtle cranial movements were essential for maintaining health and that disruptions in these rhythms could lead to various ailments.

Sutherland's ideas gained a following within the osteopathic profession, and he published his work on the topic in the early 20th century. John H. Craven, another influential osteopath, contributed to the development of cranial techniques and helped establish them as a part of osteopathic practice. During this time, cranial osteopathy was promoted and taught in osteopathic colleges, attracting practitioners who believed in the potential therapeutic benefits of cranial manipulations.

Additionally, the theory of cranial vibrations found resonance in the holistic and natural healing movements of the early 20th century. These movements sought alternatives to conventional medicine and embraced holistic approaches to health and wellness, making them more receptive to unconventional ideas like cranial rhythms.

**Disproof of the Theory**

The theory of cranial vibrations began to face skepticism and scrutiny as the medical field advanced and evidence-based practices gained prominence. Several key factors contributed to the disproof of this theory:

**1. Scientific Skepticism**: Critics within the medical and scientific communities questioned the plausibility of subtle cranial movements playing a significant role in health. Skeptics argued that the cranial bones were fused and immobile in adulthood, making the notion of rhythmic cranial movements implausible.

**2. Lack of Empirical Evidence**: Despite the claims of practitioners of cranial osteopathy, there was a lack of robust scientific evidence to support the theory of cranial vibrations. Empirical studies failed to consistently demonstrate these subtle movements or their purported therapeutic effects.

**3. Evolution of Medical Knowledge**: As medical knowledge evolved, particularly in the fields of anatomy and physiology, there was a growing understanding of the human skull's structure and function. This understanding challenged the idea that cranial bones could move rhythmically in the manner proposed by proponents of cranial vibrations.

**4. Shift Toward Evidence-Based Medicine**: The medical field increasingly adopted evidence-based practices, emphasizing the importance of rigorous scientific research and clinical trials to evaluate medical treatments and theories. The theory of cranial vibrations, lacking empirical support, fell short of meeting these rigorous standards.

**5. Regulatory Oversight**: The emergence of regulatory

bodies, such as state medical boards and professional organizations, introduced oversight and standards for medical practice. These organizations often discouraged or disallowed the use of cranial manipulations based on unproven theories.

## The Disproof of the Theory in Modern Times

In contemporary medicine, the theory of cranial vibrations is largely discredited and considered a pseudoscience. It is not recognized as a valid medical concept or treatment modality by mainstream medical organizations and institutions.

Cranial osteopathy, which was once associated with the theory of cranial vibrations, has evolved over time. It has shifted its focus away from cranial rhythms and movements and towards evidence-based practices that emphasize a holistic approach to patient care. In modern osteopathy, cranial techniques are typically used sparingly and within a broader context of evidence-based treatment.

In conclusion, the theory of cranial vibrations, which proposed that subtle rhythmic movements of the cranial bones played a vital role in health, emerged during a period of exploration within the medical field in the late 19th and early 20th centuries. While it attracted some proponents, including figures within the osteopathic and holistic medicine communities, the theory faced skepticism and scrutiny from the scientific and medical establishments. The lack of empirical evidence, scientific skepticism, and the shift toward evidence-based medicine contributed to the gradual disproof of the theory. In modern times, the theory of cranial vibrations is widely discredited and considered pseudoscientific, reflecting the evolution of medical knowledge and standards.

# Cocaine Toothache Drops

The theory of using cocaine toothache drops as a remedy for dental pain is a historical medical concept that showcases the evolution of medical practices and the sometimes misguided beliefs of the past. This theory proposed that applying cocaine to the affected tooth or gum area could provide relief from toothaches. It is a striking example of how medical knowledge and practices have changed over time, as well as the dangers associated with the use of a potent drug like cocaine for medical purposes.

**Origins of the Theory**

The theory of using cocaine for toothache relief can be traced back to the late 19th century. This period was marked by the growing popularity of cocaine and its applications in various medical contexts. Cocaine, derived from the coca plant native to South America, had been used by indigenous peoples for centuries for its stimulating and analgesic properties.

The isolation and purification of cocaine as a potent alkaloid by German chemist Albert Niemann in 1859 marked a turning point in the substance's medical use. This discovery led to the exploration of cocaine's potential applications in Western medicine.

**Prominent Believers in the Theory**

One of the earliest recorded uses of cocaine for dental pain relief can be attributed to Carl Koller, an Austrian ophthalmologist. In 1884, Koller conducted experiments that demonstrated the local anesthetic properties of cocaine when applied to the eye's surface. This groundbreaking discovery not only revolutionized ophthalmology but also piqued the interest of the dental community.

During this period, cocaine was perceived as a wonder drug with remarkable pain-relieving properties. Dentists began to explore its potential for alleviating toothaches and other oral discomforts. Cocaine-based solutions and preparations, including toothache drops, gained popularity among both dental professionals and the general public.

**The Disproof of the Theory**

The theory of using cocaine toothache drops eventually faced disproof and condemnation due to several critical factors:

**1. Growing Awareness of Cocaine's Dangers**: While cocaine was initially hailed as a medical breakthrough, it did not take long for its addictive and harmful nature to become apparent. Cocaine addiction, characterized by severe physical and psychological effects, began to surface among users. The drug's euphoric effects, which contributed to its recreational use, were also associated with significant risks.

**2. Lack of Long-Term Relief**: Despite the initial numbing effect provided by cocaine, it became evident that its use for dental pain relief offered only temporary relief and did not address the underlying causes of toothaches or dental issues. The absence of lasting benefits made its use less attractive as a dental remedy.

**3. Development of Safer Alternatives**: The disproof of the theory was hastened by the development of safer and more effective local anesthetics for dental procedures. Novocaine, introduced in the early 20th century, offered a safer alternative for pain relief during dental work. Unlike cocaine, Novocaine did not carry the same risk of addiction and had a more predictable and manageable profile for dental use.

**4. Medical and Public Awareness**: As awareness grew regarding the dangers of cocaine addiction and misuse, both the medical community and the public began to distance themselves from the drug. Cocaine was increasingly viewed as a hazardous substance rather than a medical panacea.

**When the Theory Was Disproved**

The disproof of the theory of using cocaine toothache drops occurred progressively throughout the early 20th century. Several key events and developments contributed to the decline of cocaine's popularity in dentistry:

**1. Early 20th Century**: As Novocaine and other safer local anesthetics were introduced and gained acceptance within the dental community, the use of cocaine for dental pain relief began to decline. Dentists and physicians recognized the potential for addiction and harmful side effects associated with cocaine, leading to a shift away from its use.

**2. Regulatory Measures**: The dangers of cocaine became more widely recognized, prompting regulatory measures and restrictions on its use. Government authorities and medical associations took steps to control the distribution and prescription of cocaine-containing products, further limiting its availability.

**3. Cultural Shifts**: The early 20th century also saw a

cultural shift away from the use of cocaine for recreational or medical purposes. Public awareness campaigns and educational efforts emphasized the risks associated with cocaine use, contributing to its declining popularity.

By the mid-20th century, the use of cocaine for dental pain relief had largely disappeared from mainstream dental practice. Dentists and healthcare professionals had adopted safer and more effective alternatives, and the dangers of cocaine use were widely acknowledged.

In summary, the theory of using cocaine toothache drops, which originated in the late 19th century, was based on the belief that cocaine could provide relief from dental pain. It gained popularity during a period when cocaine was viewed as a medical wonder drug with pain-relieving properties. However, the theory was eventually disproved due to growing awareness of cocaine's dangers, its lack of long-term relief, the development of safer alternatives, and regulatory measures. The decline of cocaine's use in dentistry and medicine occurred progressively throughout the early to mid-20th century, reflecting a shift in medical practices and the recognition of cocaine's risks.

# Phrenomagnetism

Phreno-magnetism, a rather curious and now largely forgotten theory, emerged during the 19th century as a fusion of two distinct but popular pseudosciences: phrenology and magnetism. This theory proposed a connection between the bumps and contours of the skull, as advocated by phrenology, and the influence of magnetic forces on an individual's mental faculties. Although it might seem peculiar today, phreno-magnetism had its proponents and briefly thrived during a time when pseudoscientific beliefs held considerable sway. In this exploration, we will delve into the essence of phreno-magnetism, its historical origins, its early proponents, the influential individuals who believed in its validity, the process by which it was ultimately disproved, and the timeline of its discrediting.

## The Essence of Phreno-Magnetism

Phreno-magnetism was a theory that aimed to combine the principles of phrenology and magnetism. Phrenology, a pseudoscience developed by Franz Joseph Gall, posited that the shape and size of the cranial bumps on an individual's skull were indicative of their mental faculties and character traits. Magnetism, on the other hand, was the belief in the existence of an invisible magnetic fluid that could influence human health and behavior. Phreno-magnetism sought to establish a connection between these two ideas, suggesting that the application of magnets to specific areas of the skull could influence a person's mental attributes, as determined by phrenology.

## Origins of Phreno-Magnetism

Phreno-magnetism emerged in the mid-19th century, a period marked by an increasing interest in pseudoscientific and paranormal phenomena.

**John Bovee Dods:** The American preacher and lecturer John Bovee Dods is credited with popularizing phreno-magnetism. Dods was known for his involvement in various unconventional and metaphysical practices, and he saw potential in merging the ideas of phrenology and magnetism.

## Proponents of Phreno-Magnetism

Phreno-magnetism had proponents and gained some following during its brief period of popularity.

**John Bovee Dods:** As the chief proponent of phreno-magnetism, Dods lectured extensively on the subject and wrote about its principles in various publications.

## The Discrediting of Phreno-Magnetism

Phreno-magnetism, while briefly popular, faced criticism and ultimately faded into obscurity.

**Scientific Skepticism:** The scientific community, including many physicians and scholars, regarded phreno-magnetism with skepticism. Critics pointed out the lack of empirical evidence to support the claims made by proponents of the theory.

**Changing Scientific Landscape:** As the 19th century progressed, the fields of psychology and neuroscience began to emerge as legitimate scientific disciplines. These developments led to a more critical examination of pseudoscientific theories like phreno-magnetism.

**Lack of Scientific Evidence:** Phreno-magnetism relied on anecdotal evidence and lacked the rigorous scientific testing and validation that genuine scientific theories undergo.

## When Phreno-Magnetism Was Disproved

Phreno-magnetism gradually lost credibility and faded from the public consciousness.

**Mid-19th Century:** The theory gained some attention and followers during the mid-19th century but failed to gain wide acceptance within the scientific community.

**Late 19th Century:** Phreno-magnetism declined in popularity as more rigorous scientific standards and methods were applied to the study of the mind and brain.

## Legacy and Insights

Phreno-magnetism serves as a historical example of the allure of pseudoscientific beliefs during a time when scientific knowledge was still emerging. It highlights the human tendency to seek unconventional explanations for complex phenomena and the importance of scientific rigor in evaluating such claims. While phreno-magnetism may seem absurd in hindsight, it should also remind us of the progress made in understanding the brain and human behavior through genuine scientific inquiry. In the end, phreno-magnetism's ultimate discrediting underscores the importance of evidence-based approaches and skepticism in the pursuit of scientific knowledge.

## *The Four Humors Theory: A Historical Lens into Ancient Medical Beliefs*

The Four Humors theory, an influential concept in the history of medicine, posited that the human body was governed by the balance of four distinct bodily fluids or humors. This theory, which originated in ancient Greece, persisted for centuries, shaping medical practices and beliefs throughout various civilizations. The theory was devised by early Greek physicians, with notable proponents like Hippocrates and Galen, and it held sway over medical thinking until the Renaissance era. In this exploration, we will delve into the essence of the Four Humors theory, its historical origins, its early proponents, the influential individuals who believed in its validity, the process by which it was eventually disproved, and the timeline of its discrediting.

### The Essence of the Four Humors Theory

The Four Humors theory postulated that the human body's health and temperament were influenced by the balance of four bodily fluids or humors: blood, phlegm, black bile, and yellow bile. Each humor was associated with specific qualities and temperament traits, and an imbalance of these humors was believed to be the root cause of illness and personality characteristics.

### Origins of the Four Humors Theory

The origins of the Four Humors theory can be traced back to ancient Greece, where early medical thinkers began to formulate theories about the body's composition and functions.

**Empedocles and Predecessors:** Early Greek philosophers like Empedocles proposed rudimentary ideas about the elements that composed the human body. These ideas laid the foundation for the development of the Four Humors theory.

**Hippocrates and the Humoral System:** Hippocrates, often referred to as the "Father of Medicine," is credited with formalizing the Four Humors theory around the 5th century BCE. He introduced the concept that an individual's health and personality were determined by the balance of the four humors.

### Proponents of the Four Humors Theory

The Four Humors theory gained widespread acceptance among physicians and scholars in the ancient and medieval worlds. Several prominent figures endorsed and expanded upon this theory.

**Galen's Influence:** Galen, a Greek physician in the 2nd century CE, played a pivotal role in popularizing the Four Humors theory. His writings and teachings, which integrated aspects of Hippocratic medicine with his own observations, had a profound influence on medieval and Renaissance medicine.

**Arabic and Medieval Scholars:** The Four Humors theory was embraced and further developed by scholars in the Arab world during the medieval period. This dissemination of knowledge helped maintain the theory's prominence in both the East and West.

### The Decline of the Four Humors Theory: Emergence of Modern Medicine

The credibility of the Four Humors theory began to wane as the scientific method gained prominence, and new discoveries contradicted its principles. Several key developments contributed to the theory's eventual discrediting.

**Advances in Anatomy:** Renaissance-era anatomists made significant strides in understanding the human body's structure and functions. These discoveries challenged the simplistic anatomical framework upon which the Four Humors theory was built.

**William Harvey's Circulation of Blood (17th Century):** William Harvey's groundbreaking work on the circulation of blood in the early 17th century provided a clear, evidence-based understanding of cardiovascular physiology, which contradicted the humoral concept of blood.

**Microscopy and Microbiology (17th-18th Centuries):** The invention of the microscope and the subsequent discoveries in microbiology revealed the existence of microorganisms and cellular structures, reshaping perceptions of disease causation and bodily functions.

**When the Four Humors Theory Was Disproved**

The discrediting of the Four Humors theory occurred over a span of centuries, marked by a gradual shift from humoral explanations of health and illness to the emergence of modern medical science.

**Renaissance Period:** The Renaissance era witnessed a gradual decline in the dominance of the Four Humors theory, as anatomical discoveries and new scientific methods began to challenge its premises.

**17th-18th Centuries:** The work of scientists like William Harvey, Robert Boyle, and Antonie van Leeuwenhoek in the 17th and 18th centuries laid the groundwork for a more evidence-based understanding of the human body.

**19th Century:** The 19th century saw significant advancements in medical science, including the development of modern pathology and the understanding of infectious diseases, which further undermined the Four Humors theory.

**Legacy and Insights**

The Four Humors theory serves as a historical testament to the evolving nature of medical knowledge. While it once held sway over medical practices for centuries, it was ultimately replaced by more accurate and evidence-based understandings of the human body. The theory's legacy lies in its role as a stepping stone in the development of modern medicine, shedding light on how our understanding of health and disease has evolved through the ages. It serves as a reminder of the importance of empirical evidence and scientific inquiry in shaping our comprehension of the world around us.

*Bloodletting: The Historical Practice and Its Demise*

Bloodletting, an ancient medical practice, is a testament to the evolution of medical knowledge and the sometimes harmful consequences of tradition. This practice, which involved the deliberate removal of blood from a patient's body, was rooted in the belief that it could restore health by balancing the body's humors. In this exploration, we will delve into the history of bloodletting, its origins, its proponents, the eventual debunking of its principles, and the timeline of its decline.

**The Essence of Bloodletting**

Bloodletting, also known as phlebotomy, was based on

the theory of the four humors, a concept inherited from ancient Greek and Roman medicine. According to this theory, the body's health depended on a balance of four bodily fluids: blood, phlegm, black bile, and yellow bile. Illness was thought to result from an excess or imbalance of these humors. Bloodletting aimed to rectify this imbalance by removing excess blood.

### Origins of Bloodletting

The origins of bloodletting can be traced back to ancient civilizations:

**Ancient Egypt:** The practice of bloodletting dates as far back as 1000 BCE in ancient Egypt. Egyptians used tools like lancets and leeches to withdraw blood, believing it could purify the body and treat various ailments.

**Ancient Greece:** Hippocrates, often considered the father of modern medicine, advocated for bloodletting as a means to balance the humors. His influence spread throughout the Mediterranean, perpetuating the practice.

### Proponents and Believers

Bloodletting was endorsed and practiced by numerous historical figures, including:

**Galen (129-200 CE):** Galen, a Greek physician in the Roman Empire, played a significant role in popularizing bloodletting. His writings and teachings promoted the practice as a remedy for various diseases.

**Medieval and Renaissance Physicians:** Bloodletting remained a standard medical treatment throughout the Middle Ages and the Renaissance. Physicians like Avicenna and Andreas Vesalius continued to advocate for its use.

**George Washington:** Even notable figures like the first President of the United States, George Washington, fell victim to bloodletting. In December 1799, he was bled extensively as treatment for a severe throat infection, ultimately leading to his death.

### The Demise of Bloodletting

The decline of bloodletting was a gradual process that unfolded over several centuries, marked by key developments:

**Empirical Observations:** As medical knowledge advanced, some physicians began to question the effectiveness of bloodletting. Empirical observations and the lack of consistent positive outcomes raised doubts about the practice.

**Advancements in Medical Science:** The 19th century witnessed significant advancements in medical science. The discovery of bacteria, the development of the germ theory of disease, and the understanding of the circulatory system provided a more accurate understanding of illness and rendered bloodletting obsolete.

**Introduction of Evidence-Based Medicine:** The rise of evidence-based medicine in the late 19th and early 20th centuries emphasized the importance of clinical trials and empirical evidence. Bloodletting could not withstand the scrutiny of scientific rigor.

### When Bloodletting Got Disproved

Bloodletting gradually fell out of favor as scientific understanding advanced, but it can be pinpointed to the late 19th and early 20th centuries when it was officially discredited:

**Late 19th Century:** By this time, prominent physicians and scientists had begun to question the efficacy of bloodletting. The advent of evidence-based medicine and the development of more effective treatments hastened its decline.

**Early 20th Century:** With the publication of groundbreaking research and the widespread adoption of evidence-based practices, bloodletting became increasingly rare in medical practice. Its disproof was formalized as physicians abandoned it in favor of more effective treatments.

In conclusion, bloodletting, a practice rooted in the ancient theory of the four humors, was once widely accepted as a medical treatment. It persisted for centuries and was promoted by influential figures in the medical field. However, as medical science advanced and empirical evidence became the cornerstone of medicine, bloodletting gradually lost its credibility. The decline of this practice can be attributed to the skepticism of physicians, the growth of evidence-based medicine, and the development of a more accurate understanding of disease and the circulatory system. Bloodletting was eventually discredited in the late 19th and early 20th centuries, marking a significant turning point in the history of medicine.

# Dousing with Mercury

The theory of dousing with mercury is a historical medical practice that exemplifies the evolution of medical knowledge and the sometimes perilous treatments that were administered in the past. This theory proposed that mercury, a highly toxic heavy metal, could be used as a therapeutic agent to treat various ailments. It is a striking example of how misguided beliefs, lack of scientific understanding, and the absence of regulatory oversight once allowed dangerous treatments to be administered to patients.

**Origins of the Theory**

The theory of dousing with mercury has its origins in ancient times, where mercury was regarded as a mysterious and potent substance. Mercury, a heavy, silvery metal, has been known to humanity for thousands of years and was often associated with alchemical practices. Due to its unique physical properties, including its liquid state at room temperature, it gained a reputation as a magical or mystical substance.

However, the medical use of mercury gained prominence during the late Middle Ages and the Renaissance period in Europe. During this time, mercury was believed to possess therapeutic properties, and it was used to treat a wide range of ailments, including syphilis, smallpox, and various neurological disorders.

**Prominent Believers in the Theory**

Many historical figures, including physicians and scientists, believed in the therapeutic properties of mercury during different periods of history. Some notable proponents of mercury-based treatments included:

**Paracelsus (1493-1541)**: Paracelsus, a Swiss physician, alchemist, and philosopher, was a prominent figure in the history of medicine. He believed in the concept of using "like to treat like" and advocated for the use of mercury in certain medical treatments.

**Sir William Osler (1849-1919)**: Sir William Osler, a renowned Canadian physician, was a pioneer in modern medicine. However, he recommended mercury-based treatments for syphilis, reflecting the prevailing medical beliefs of his time.

**Various Historical Medical Texts**: Mercury-based treatments were documented in numerous medical texts throughout history. These texts often prescribed mercury-containing compounds for a wide range of conditions.

**The Disproof of the Theory**

The theory of dousing with mercury eventually faced disproof and condemnation due to several critical factors:

**Harmful Effects**: Over time, it became evident that mercury exposure was associated with severe and often irreversible health consequences. Mercury poisoning, characterized by a range of symptoms including tremors, cognitive impairment, and organ damage, became widely recognized as a significant risk associated with mercury-based treatments.

**Advancements in Medical Knowledge**: As medical

knowledge advanced, the harmful effects of mercury became better understood. The toxic nature of mercury and its potential to cause harm to the human body led to a reassessment of its medical use.

**Alternative Treatments**: The development and adoption of alternative, safer, and more effective treatments for the conditions that had been treated with mercury led to a decline in its use. For example, the discovery of antibiotics revolutionized the treatment of syphilis, making mercury-based treatments obsolete.

### When the Theory Was Disproved

The disproof of the theory of dousing with mercury occurred progressively throughout the 19th and 20th centuries. Several key factors contributed to the decline and eventual disproof of this dangerous medical practice:

**19th Century**: The 19th century witnessed a growing understanding of the toxic properties of mercury and its harmful effects on the human body. Medical professionals and scientists began to recognize the dangers associated with mercury-based treatments.

**Shift Toward Evidence-Based Medicine**: As medicine embraced evidence-based practices, the use of mercury without substantial scientific evidence or clinical trials became increasingly untenable. The lack of empirical support for mercury-based treatments contributed to their discrediting.

**Regulatory Measures**: Governments and medical authorities introduced regulatory measures to control the use of toxic substances like mercury in medical practice. These regulations aimed to protect the public from harmful treatments and promote safe and effective alternatives.

By the mid-20th century, the theory of dousing with

mercury had been unequivocally disproved and condemned by the medical community. The dangers of mercury exposure and its lack of therapeutic benefit led to its abandonment as a medical treatment.

In summary, the theory of dousing with mercury, which proposed the use of highly toxic mercury as a therapeutic agent, has its origins in ancient and medieval medical practices. It gained prominence during the Renaissance and later periods, with notable historical figures endorsing its use. However, the disproof of the theory occurred as scientific understanding advanced, and the harmful effects of mercury became apparent. Advancements in medical knowledge, the development of alternative treatments, and regulatory measures led to the abandonment of mercury-based therapies. The theory of dousing with mercury serves as a stark reminder of the evolution of medical knowledge and the importance of evidence-based medicine in safeguarding patient health.

# Urine Therapy

The theory of urine therapy is a controversial and unconventional belief that advocates the consumption or application of one's own urine for various health and wellness benefits. This theory, rooted in alternative medicine and folk remedies, has its origins in ancient traditions and has been promoted by various individuals over the years. While proponents of urine therapy claim a wide range of health benefits, the scientific and medical communities have consistently rejected these assertions, viewing the practice as unproven and potentially harmful.

**Origins of the Theory**

The practice of urine therapy has ancient origins and can be traced back to several traditional healing systems, including Ayurveda and traditional Chinese medicine. In Ayurveda, urine is considered a waste product that can be utilized for medicinal purposes. These historical practices were based on the belief that urine contains valuable substances and energies that can promote healing and balance within the body.

The modern revival of urine therapy can be attributed to various individuals who popularized the practice in the 20th century. One notable figure is John W. Armstrong, a British naturopath, who authored the book "The Water of

Life: A Treatise on Urine Therapy" in 1944. Armstrong's book espoused the therapeutic potential of urine and claimed that urine therapy could cure a wide range of ailments.

**Prominent Believers in the Theory**

Urine therapy has attracted proponents from alternative medicine and natural healing communities. These proponents often argue that urine contains valuable nutrients, hormones, and antibodies that can have healing effects on the body. Some of the claims made by proponents of urine therapy include its use for skin conditions, as a nutritional supplement, and even as a cancer treatment.

While there have been various individuals who have promoted urine therapy, it's important to note that this practice has never gained widespread acceptance within the scientific or medical communities. Proponents often cite anecdotal evidence and personal testimonials to support their claims, but such evidence falls far short of the rigorous scientific standards required to validate medical treatments.

**The Disproof of the Theory**

The theory of urine therapy has been consistently disproved and rejected by the scientific and medical communities for several reasons:

**1. Lack of Scientific Evidence**: Despite the claims made by proponents of urine therapy, there is a conspicuous absence of robust scientific research or clinical trials supporting its efficacy. Scientific studies have not provided conclusive evidence that urine contains therapeutic substances or that the practice of urine therapy offers any measurable health benefits.

**2. The Role of Kidneys**: The kidneys are responsible for filtering waste products and toxins from the bloodstream, which are then excreted in urine. Consuming or applying urine involves reintroducing these waste products back into the body, which contradicts the body's natural processes for waste elimination.

**3. Potential Health Risks**: Urine is not a sterile fluid, and it may contain harmful bacteria or pathogens that can pose health risks when ingested or applied to the skin. Furthermore, the consumption of one's own urine could potentially lead to electrolyte imbalances or other health complications.

**4. Alternative and Proven Therapies**: Modern medicine offers well-established, evidence-based treatments and therapies for various health conditions. When compared to these scientifically validated approaches, urine therapy lacks a solid foundation of evidence and carries potential risks.

**When the Theory Was Disproved**

The theory of urine therapy has been discredited and rejected by the scientific and medical communities throughout its modern history. While proponents of urine therapy continue to advocate for its use, their claims have not gained scientific legitimacy or acceptance.

The disproof of the theory has been an ongoing process, as the scientific community has consistently assessed and scrutinized the claims associated with urine therapy. The lack of scientific evidence and the potential health risks associated with the practice have contributed to its rejection by mainstream medicine.

In contemporary times, urine therapy remains a fringe belief and is not recognized as a valid medical treatment. It has not gained acceptance within the scientific or medical communities, and it is not supported by evidence-based practices.

In summary, the theory of urine therapy advocates the consumption or application of one's own urine for various health benefits, rooted in ancient healing traditions and promoted by individuals in the 20th century. While proponents claim a wide range of therapeutic effects, the practice has consistently been disapproved and rejected by the scientific and medical communities due to the lack of scientific evidence, potential health risks, and the availability of established, evidence-based medical treatments. Urine therapy remains a fringe belief and is not recognized as a valid medical practice in contemporary times.

# Testicle Transplants for Virility

The theory of using testicle transplants for virility is a historical and pseudoscientific belief that proposed the surgical transplantation of testicles from young individuals into older men as a means to restore or enhance virility, vitality, and youthfulness. This theory, rooted in the late 19th and early 20th centuries, was based on the mistaken idea that testicles contained vital substances that could rejuvenate the recipient. It is a prime example of how medical practices and beliefs have evolved over time, moving away from unproven and potentially harmful treatments toward evidence-based medicine.

**Origins of the Theory**

The theory of testicle transplants for virility has its origins in the late 19th and early 20th centuries, a period when medical understanding was still evolving, and the mechanisms behind many physiological processes remained unknown. During this time, there was a fascination with the concept of rejuvenation and the search for ways to slow down the aging process.

One of the key figures associated with the promotion of testicle transplants was Dr. Serge Voronoff, a Russian-born French surgeon. Voronoff gained notoriety for his work in grafting monkey testicles into humans, believing that the procedure would enhance vitality and virility. His theories and experiments contributed to the popularization of the idea that testicle transplants could rejuvenate aging men.

## Prominent Believers in the Theory

The theory of testicle transplants for virility found proponents among a subset of physicians, scientists, and individuals seeking to combat the effects of aging. Some notable figures who believed in the theory include:

**Dr. Serge Voronoff**: As the pioneer of testicle transplantation, Voronoff was a fervent believer in the rejuvenating effects of the procedure. He performed numerous monkey-to-human testicle transplants and claimed that the grafts could boost energy, strength, and overall vitality.

**Dr. John R. Brinkley**: An American physician, Brinkley, was infamous for his "goat gland" transplant surgeries in the early 20th century. He believed that transplanting goat testicles into men could enhance virility and cure various ailments. While not the same as human testicle transplants, this idea shared a similar pseudoscientific foundation.

## The Story of How the Theory Eventually Got Disproved

The theory of testicle transplants for virility was gradually discredited and abandoned for several key reasons:

**1. Lack of Scientific Evidence**: Despite the claims made by proponents like Voronoff and Brinkley, there was a glaring absence of robust scientific evidence to support the idea that testicle transplants could rejuvenate or enhance virility. The purported benefits were largely anecdotal and lacked rigorous scientific validation.

**2. Ethical and Surgical Challenges**: The procedures

themselves posed significant ethical dilemmas and surgical challenges. Transplanting human testicles from one individual to another was a complex and risky surgery that often resulted in complications, including infection and graft rejection.

**3. Advances in Endocrinology**: As the field of endocrinology advanced, scientists gained a better understanding of hormones and their role in physiological processes. It became apparent that the hormones responsible for sexual development and function were primarily produced by the testes, but simply transplanting testicles did not necessarily restore hormonal balance or enhance virility.

**4. Ethical Concerns**: The ethical concerns surrounding the use of human testicle transplants, especially from younger donors to older recipients, raised questions about consent and bodily autonomy. This contributed to a more critical examination of the practice.

### When the Theory Was Disproved

The theory of using testicle transplants for virility gradually lost credibility and was discredited from the early to mid-20th century:

**Early 20th Century**: Voronoff and Brinkley's experiments garnered attention and controversy in the early 20th century. However, as scientific understanding advanced, and the lack of substantial evidence became apparent, the theory began to face criticism and skepticism.

**Mid-20th Century**: By the mid-20th century, the use of testicle transplants for virility had largely been abandoned within the medical community. Regulatory authorities began to take action against practitioners promoting these unproven and ethically questionable procedures.

**Contemporary Times**: In contemporary medicine, the theory of testicle transplants for virility is viewed as a historical curiosity and a stark example of pseudoscience. It serves as a reminder of the importance of evidence-based medicine and ethical considerations in medical practice.

In summary, the theory of using testicle transplants for virility originated in the late 19th and early 20th centuries, driven by the misguided belief that such transplants could rejuvenate aging men. Figures like Dr. Serge Voronoff and Dr. John R. Brinkley promoted this theory, but it was gradually disproved due to a lack of scientific evidence, ethical concerns, surgical challenges, and advances in endocrinology. By the mid-20th century, the practice of testicle transplants for virility had largely disappeared from medical practice, marking the decline of a pseudoscientific and ethically problematic concept.

# Heroin as a Cough Suppressant

The theory of using heroin as a cough suppressant is a historical medical practice that highlights the evolving understanding of pharmaceuticals and the sometimes misguided use of potent substances for medicinal purposes. This theory suggested that heroin, a synthetic opioid derived from morphine, could effectively relieve coughing and other respiratory symptoms. It is a stark example of how medical knowledge has advanced over time and the risks associated with the use of powerful drugs in the absence of rigorous scientific evaluation.

**Origins of the Theory**

The use of heroin as a cough suppressant has its roots in the late 19th and early 20th centuries when pharmaceutical companies were actively developing and marketing new medications. Heroin itself was first synthesized by chemist C.R. Alder Wright in 1874 but remained largely unexplored until it was re-synthesized by Bayer, a German pharmaceutical company, in 1897. Bayer initially marketed heroin as a non-addictive alternative to morphine and as a treatment for coughs and pain.

**Prominent Believers in the Theory**

During the early 20th century, the theory of using heroin as a cough suppressant gained prominence and was endorsed by pharmaceutical companies, physicians, and even the medical community. Some notable individuals and entities that promoted this theory include:

**Pharmaceutical Companies**: Bayer, in particular, played a significant role in promoting heroin as a cough remedy. They marketed the drug under the brand name "Heroin" and claimed it was a safe and effective treatment for coughs, colds, and respiratory issues.

**Medical Practitioners**: Many physicians and healthcare providers prescribed heroin-based cough syrups and medications to their patients. The belief in the drug's efficacy was widespread among medical professionals of the time.

**Public Perception**: The general public also had a positive perception of heroin as a cough remedy due to extensive advertising and endorsements by reputable pharmaceutical companies.

**The Story of How the Theory Eventually Got Disproved**

The theory of using heroin as a cough suppressant gradually fell out of favor and was eventually disproved due to several significant factors:

**1. Addictive Nature of Heroin**: It became evident that heroin was not the safe and non-addictive drug that it was initially marketed as. Individuals who used heroin for medicinal purposes were at risk of developing addiction and physical dependence on the drug. The realization of heroin's addictive potential raised serious concerns about its widespread use.

**2. Increased Regulation**: As concerns grew over the misuse and addictive nature of heroin, regulatory authorities began to impose stricter controls on its production, distribution, and prescription. Many countries implemented measures to restrict access to heroin-based medications.

**3. Emergence of Safer Alternatives**: The development of safer and more effective cough suppressants, such as codeine and dextromethorphan, offered viable alternatives to heroin. These newer medications provided cough relief without the risk of addiction associated with heroin.

**4. Scientific Understanding of Opioids**: As scientific knowledge advanced, there was a better understanding of the pharmacology of opioids and their potential for harm. Medical professionals began to recognize the risks and side effects associated with opioid use, including respiratory depression.

**When the Theory Was Disproved**

The theory of using heroin as a cough suppressant was progressively disproved and fell out of favor from the early to mid-20th century:

**Early 20th Century**: During the early 1900s, concerns about heroin addiction and its potential for misuse began to emerge. This led to increased scrutiny and regulation of heroin-based medications.

**Mid-20th Century**: By the mid-20th century, the use of heroin for cough suppression and other medical purposes had largely been abandoned. Regulatory measures and the emergence of safer alternatives contributed to the decline of heroin-based medications in the medical field.

**Contemporary Times**: In contemporary medicine, the use of heroin as a cough suppressant is viewed as a historical curiosity and a stark example of the evolving understanding of pharmaceuticals and their risks. Heroin is now recognized as a highly addictive and dangerous drug, and its use in medical practice has been replaced by safer and more effective alternatives.

In summary, the theory of using heroin as a cough suppressant originated in the late 19th and early 20th centuries when pharmaceutical companies marketed it as a non-addictive remedy for respiratory symptoms. The theory was widely promoted and endorsed by pharmaceutical companies, physicians, and the public. However, it was gradually disproved due to the addictive nature of heroin, increased regulation, the emergence of safer alternatives, and a better scientific understanding of opioids. By the mid-20th century, the use of heroin for cough suppression had largely ceased, and it is now recognized as a dangerous and addictive substance in contemporary medicine.

# Electric Bathing

The theory of electric bathing was an unconventional and pseudoscientific belief that promoted the therapeutic use of electrically charged water for various health benefits. This theory, rooted in the late 18th and early 19th centuries, claimed that immersing oneself in electrically charged water could cure various ailments and promote general well-being. It is an example of how scientific understanding and medical practices have evolved over time, as this once-popular concept eventually gave way to evidence-based medicine.

**Origins of the Theory**

The theory of electric bathing has its origins in the late 18th century when electricity was a subject of great scientific curiosity and experimentation. During this period, electricity was not fully understood, and it held an aura of mystery and fascination. Medical practitioners and inventors began to explore the potential therapeutic applications of electricity, leading to the development of various electrotherapy devices and treatments.

One of the key figures associated with the early promotion of electric bathing was Dr. Giovanni Aldini, an Italian physicist and professor. Aldini conducted experiments and demonstrations using electricity on both animals and humans, which garnered significant attention in Europe. His work laid the groundwork for the belief in the healing properties of electrically charged water.

**Prominent Believers in the Theory**

The theory of electric bathing found proponents among both medical practitioners and the general public during the late 18th and early 19th centuries. Some notable believers and advocates of electric bathing included:

**Dr. Giovanni Aldini**: As mentioned earlier, Aldini played a significant role in popularizing the therapeutic use of electricity, including electric bathing. His experiments and public demonstrations drew considerable interest and contributed to the theory's propagation.

**Dr. John Wesley**: The theory of electric bathing was also endorsed by Dr. John Wesley, an English cleric, and one of the founders of Methodism. Wesley believed in the health benefits of electric baths and promoted their use in his writings.

**Various Inventors**: The early 19th century saw the invention and marketing of electric bath apparatus by individuals who claimed their devices could provide various health benefits. These devices were often marketed to the general public, promising relief from various ailments.

**The Disproof of the Theory**

The theory of electric bathing was eventually discredited and declined in popularity for several key reasons:

**1. Lack of Scientific Basis**: While electricity was a subject of fascination during the late 18th and early 19th centuries, the theory of electric bathing lacked a sound scientific basis. There was limited understanding of the physiological effects of electrically charged water on the human body, and the claims made about its health benefits were often unsubstantiated.

**2.  Rise of Evidence-Based Medicine**: As the field of medicine progressed, it increasingly embraced evidence-based practices.  Medical professionals recognized the importance of rigorous scientific research and clinical trials to validate treatments. Electric bathing did not meet these standards and could not provide credible evidence of its efficacy.

**3.  Potential Dangers**: As more was learned about the potential dangers of electricity, including the risk of electric shock, burns, and other adverse effects, the use of electricity in medical treatments came under scrutiny.  The safety of electric baths and their potential harm became a concern.

**4.  Emergence of Effective Therapies**: The disproof of the theory of electric bathing coincided with the development of more effective and evidence-based medical treatments. As medical knowledge advanced, treatments and therapies that could address specific health conditions with greater precision and safety became available.

**When the Theory Was Disproved**

The decline and disproof of the theory of electric bathing occurred progressively throughout the 19th century. As scientific understanding advanced and evidence-based medicine gained prominence, the use of electric bathing for therapeutic purposes gradually fell out of favor.

The disproof of this theory can be associated with the broader shift away from unproven and potentially harmful medical practices and towards treatments supported by scientific evidence and clinical trials.  By the mid-to-late 19th century, electric bathing had lost its credibility within the medical community and was largely viewed as a pseudoscientific and outdated concept.

In contemporary times, the theory of electric bathing is a historical curiosity, a testament to the evolving nature of medical knowledge and practices. It serves as a reminder of the importance of critical evaluation and evidence-based medicine in determining the safety and efficacy of medical treatments.

# II

# Astronomy

# The Phaeton Hypothesis

*The Phaeton Hypothesis is a scientific proposal that suggests the existence of a hypothetical planet called Phaeton and its role in the formation of the asteroid belt between Mars and Jupiter. This intriguing idea emerged in the 19th century and has since been largely dismissed by the scientific community. In this exploration, we will delve into the theory's origins, key proponents, its gradual discrediting, and when it was ultimately disproved.*

**Unraveling the Phaeton Hypothesis**

The Phaeton Hypothesis revolves around several key concepts:

**The Missing Planet:** The theory posits the existence of a planet named Phaeton, located between Mars and Jupiter. According to proponents of the hypothesis, this planet was a sizable celestial body that once orbited the Sun but no longer exists.

**Destruction of Phaeton:** The Phaeton Hypothesis suggests that Phaeton experienced a catastrophic event that led to its destruction. This event is often described as an enormous collision or explosion.

**Asteroid Belt Formation:** It is further proposed that the debris resulting from the destruction of Phaeton became the building blocks for the asteroid belt, the region of space filled with numerous small celestial objects located between the orbits of Mars and Jupiter.

**Origins of the Phaeton Hypothesis**

The Phaeton Hypothesis finds its roots in the 19th century:

**Édouard Roche:** Although not the originator of the Phaeton Hypothesis, French astronomer Édouard Roche made significant contributions to the understanding of celestial mechanics and planetary formation. His work on tidal forces and celestial dynamics indirectly influenced the development of the hypothesis.

**Supporters and Believers**

While the Phaeton Hypothesis has not garnered widespread support, it did attract some attention and notable figures who explored its implications:

**Thomas Jefferson Jackson See:** An American astronomer, See proposed the idea of a former planet's destruction leading to the formation of the asteroid belt in the early 20th century. He contributed to the hypothesis's development but did not receive broad scientific endorsement.

**Daniel Kirkwood:** An American astronomer known for his work on the asteroid belt, Kirkwood observed gaps in the distribution of asteroids that were later named Kirkwood gaps. These gaps were initially attributed to the gravitational influence of a missing planet, possibly Phaeton.

**The Demise of the Phaeton Hypothesis**

The Phaeton Hypothesis faced several challenges that contributed to its gradual discrediting:

**Lack of Evidence:** One of the primary issues with the hypothesis was the lack of concrete evidence supporting the existence of Phaeton or the catastrophic event that led to its destruction. Observations and measurements did not align with the theory's predictions.

**Advancements in Celestial Mechanics:** As astronomical knowledge and computational tools advanced, scientists gained a better understanding of celestial mechanics. The gravitational dynamics of the solar system could be explained without invoking the presence of a missing planet.

**Discovery of Asteroid Families:** The discovery of asteroid families within the asteroid belt provided an alternative explanation for the distribution of asteroids. These families could be attributed to the gravitational interactions and collisions among asteroids themselves, rather than the influence of a former planet.

**Improved Telescopes and Space Missions:** Technological advancements, such as more powerful telescopes and space missions to study asteroids, provided detailed information about the composition, orbits, and characteristics of asteroids, further undermining the need for a missing planet.

## When the Phaeton Hypothesis Got Disproved

The Phaeton Hypothesis gradually lost credibility and can be considered largely disproved by the mid-to-late 20th century. As scientific understanding of the solar system improved and alternative explanations emerged, the idea of a former planet called Phaeton became less tenable.

In conclusion, the Phaeton Hypothesis proposed the existence of a hypothetical planet named Phaeton and its role in the formation of the asteroid belt. It originated in the 19th century and gained some attention from astronomers like Thomas Jefferson Jackson See and Daniel Kirkwood. However, the hypothesis faced significant challenges, including a lack of evidence and advancements in celestial mechanics and technology. As a result, it lost support within the scientific community and can be considered largely disproved by the mid-to-late 20th century.

# The Nebular Hypothesis

The nebular hypothesis is a foundational concept in the field of astronomy and cosmology. It provides a comprehensive explanation for the formation of our solar system and, by extension, other planetary systems in the universe. In this exploration, we will delve into the essence of the nebular hypothesis, its historical origins, its proponents, the evidence that supported it, and how it continues to shape our understanding of the cosmos.

**The Essence of the Nebular Hypothesis**

At its core, the nebular hypothesis proposes that stars and planetary systems form from the gravitational collapse of a giant cloud of gas and dust in space, known as a nebula. This cloud undergoes a series of transformations, ultimately leading to the creation of stars, planets, moons, and other celestial bodies. The key stages of this process can be summarized as follows:

**Formation of a Nebula:** The process begins with the formation of a massive cloud of gas and dust in space. This nebula may be triggered by various factors, such as the shockwave from a nearby supernova or the gravitational perturbation of another passing celestial object.

**Gravitational Collapse:** Over time, the nebula's own

gravitational attraction causes it to collapse inward. As the cloud contracts, it begins to spin due to the conservation of angular momentum, forming a spinning disk-shaped structure.

**Protostar Formation:** In the center of this spinning disk, material accumulates, and the density and temperature increase. Eventually, a protostar forms at the core of the collapsing cloud. A protostar is a young star in the early stages of its formation.

**Accretion of Planetary Material:** As the protostar continues to grow, its gravitational influence extends throughout the disk. In regions farther from the protostar, solid particles collide and stick together, forming planetesimals. These planetesimals, in turn, collide and merge to form planets.

**Formation of a Planetary System:** The remaining material in the disk forms a range of celestial objects, including moons, asteroids, and comets. Over time, the protostar reaches a stable state and becomes a full-fledged star. The planetary system, with its newly formed planets, orbits the star.

### Origins of the Nebular Hypothesis

The origins of the nebular hypothesis can be traced back to the late 18th century and the work of the French mathematician and astronomer Pierre-Simon Laplace. In 1796, Laplace published his seminal work, "Exposition du système du monde" (Exposition of the System of the World), in which he presented his ideas on the formation of the solar system.

### Proponents and Believers

The nebular hypothesis gained widespread acceptance and has been embraced by numerous scientists and astronomers over the years. Some of the notable proponents and supporters of the nebular hypothesis include:

**Pierre-Simon Laplace:** As the original proponent of the hypothesis, Laplace played a foundational role in shaping our understanding of the formation of planetary systems.

**Sir William Herschel:** The renowned British astronomer Sir William Herschel, known for his discovery of Uranus, supported the nebular hypothesis. His observations of the universe were consistent with its predictions.

**Immanuel Kant:** The German philosopher Immanuel Kant independently proposed a similar hypothesis for the formation of planetary systems, which closely aligned with Laplace's ideas.

**Victor Safronov:** In the mid-20th century, Russian astrophysicist Victor Safronov provided mathematical support for the nebular hypothesis through his studies on the dynamics of planetary formation.

**Contemporary Astronomers:** Today, the nebular hypothesis remains a cornerstone of planetary science and astrophysics. Contemporary astronomers continue to refine and expand upon the theory, using advanced techniques and technologies to explore planetary formation in other star systems.

### Evidence and Confirmation

Over the years, extensive evidence has supported the nebular hypothesis and confirmed its validity. Some of the key lines of evidence include:

**Solar System Composition:** The composition and distribution of elements and compounds within our solar system align with the predictions of the nebular hypothesis. It explains why the inner planets are rocky and dense, while the outer planets are gas giants.

**Observations of Young Stars:** Astronomers have observed protoplanetary disks around young stars, providing direct evidence of the early stages of planetary formation.

**Orbital Characteristics:** The orbital characteristics of planets, such as their distances from the Sun and their orbital planes, can be explained by the nebular hypothesis.

**Computer Simulations:** Modern computer simulations of planetary formation consistently produce outcomes that are consistent with the predictions of the nebular hypothesis.

### The Nebular Hypothesis Today

The nebular hypothesis remains a fundamental concept in astrophysics and planetary science. It has been instrumental in shaping our understanding of how planetary systems, including our own solar system, came into existence. Additionally, ongoing research, space exploration missions, and advancements in observational techniques continue to refine and expand our knowledge of planetary formation processes in other star systems.

In summary, the nebular hypothesis is a pivotal theory that explains the formation of stars and planetary systems from the gravitational collapse of nebular clouds. It originated in the late 18th century with Pierre-Simon Laplace and has since gained wide acceptance in the scientific community. The hypothesis is supported by a wealth of evidence, and ongoing research continues to enhance our understanding of planetary formation. It stands as a testament to the enduring power of scientific inquiry and discovery in unraveling the mysteries of the cosmos

# The Theory of Vulcan

The theory of Vulcan postulated the existence of an unseen planet located closer to the Sun than Mercury. Its creation stemmed from the inability of astronomers to explain discrepancies in Mercury's orbit using Newtonian physics alone. Specifically, Mercury's perihelion, the point in its orbit closest to the Sun, appeared to shift more than expected over time. This unexplained orbital precession led to the idea that another massive body, Vulcan, might be tugging on Mercury.

**Origins of the Theory:**

The concept of Vulcan was introduced in the early 19th century, with the first mention usually attributed to French mathematician Urbain Le Verrier. Le Verrier, renowned for his prediction of Neptune's existence based on deviations in Uranus's orbit, was a prominent figure in celestial mechanics. In the mid-19th century, he turned his attention to Mercury's peculiarities and began to suspect the presence of an undiscovered planet.

**Believers in Vulcan:**

Le Verrier's proposal gained traction among astronomers and the public. The idea of an unseen planet within our solar system was both captivating and plausible given Le Verrier's earlier success with Neptune. Many astronomers and scientists believed in the existence of Vulcan, hoping that its discovery would once again demonstrate the power of celestial mechanics.

**The Demise of Vulcan:**

Despite widespread belief in Vulcan's existence, its actual discovery proved elusive. Observers scoured the skies in search of the mysterious planet, but it remained hidden. As telescopes and observational techniques improved, doubts began to emerge. Some astronomers questioned whether Vulcan could really explain Mercury's orbital precession.

The ultimate demise of Vulcan came with the development of Albert Einstein's theory of General Relativity in the early 20th century. Einstein's theory provided a more accurate description of gravity than Newton's laws, and it explained Mercury's orbital precession without the need for an additional planet. According to General Relativity, the curvature of space-time around the Sun was responsible for Mercury's behavior.

**The Disproof of Vulcan:**

Einstein's General Relativity was confirmed through experimental evidence, such as the 1919 solar eclipse observations led by Sir Arthur Eddington. This confirmation of General Relativity effectively rendered Vulcan unnecessary. Scientists no longer needed an unseen planet to account for Mercury's orbit, as the new theory provided a more accurate description of the phenomenon.

**Conclusion:**

In summary, the theory of Vulcan, a hypothetical planet proposed to explain peculiarities in Mercury's orbit, originated in the early 19th century with Urbain Le Verrier. It gained widespread support among astronomers and the public but was eventually disproved with the development of Einstein's General Relativity. This theory provided a more accurate description of gravity, eliminating the need for an additional planet within our solar system. As a result, Vulcan remains a historical footnote in the evolution of our understanding of the cosmos.

# Spontaneous Generation of Planets

The theory of spontaneous generation of planets, once considered a plausible explanation for the existence of celestial bodies, has undergone a remarkable transformation over centuries. In this exploration, we will delve into the essence of this theory, its historical origins, the key figures associated with it, its believers, the gradual process of its disproof, and the eventual establishment of modern planetary formation theories.

**The Essence of the Theory of Spontaneous Generation of Planets**

The theory of spontaneous generation of planets, often referred to as the "preformationist" theory, proposed that planets and other celestial bodies came into existence spontaneously, without the need for a specific formation process. This theory posited that the universe possessed the inherent capability to generate celestial objects, much like life was believed to spontaneously generate through the theory of abiogenesis. It essentially asserted that planets appeared fully formed, as if by magic or divine creation, rather than undergoing a systematic process of formation.

**Origins of the Theory**

The roots of the theory of spontaneous generation of

planets can be traced back to ancient civilizations. Many early cultures and societies believed in creation myths that involved the spontaneous emergence of celestial bodies. These myths often incorporated gods or divine beings as the creators of planets and stars.

However, in the context of scientific thought, the theory gained traction during the Middle Ages and the Renaissance period. During this time, natural philosophy (an early form of science) was heavily influenced by religious and mystical beliefs. Scholars of the time grappled with the question of how celestial bodies came into existence, and some proposed the idea that they were spontaneously generated.

### Key Figures and Believers

While the theory of spontaneous generation of planets lacked a single originator, it was endorsed by various scholars and thinkers throughout history who were influenced by the prevailing scientific and philosophical ideas of their time. Some key figures associated with this theory include:

**Medieval Scholars:** During the Middle Ages, many medieval scholars accepted the idea of spontaneous generation, believing that celestial bodies, including planets, were created by divine intervention or through mysterious natural processes.

**Hermeticism:** The Hermetic tradition, a blend of religious and philosophical beliefs, embraced the concept of spontaneous generation as part of its mystical worldview. This influenced the thinking of some Renaissance-era scholars.

**Alchemists:** Some alchemists of the Renaissance period, who sought to transmute base metals into gold and discover the philosopher's stone, also incorporated ideas of spontaneous generation into their mystical practices.

**Preformationists:** In the 17th and 18th centuries, pre-formationist theories, which proposed that miniature fully formed beings or objects existed within larger entities, were in vogue. While not explicitly focused on planets, these theories hinted at the idea of spontaneous generation in the celestial realm.

### Gradual Disproof and Transition

The gradual disproof of the theory of spontaneous generation of planets can be attributed to several key developments:

**Telescopic Observations:** The advent of telescopes in the early 17th century allowed astronomers like Galileo Galilei and Johannes Kepler to observe celestial bodies in detail. These observations revealed that planets had distinct features, such as moons and phases, which suggested they were not spontaneously generated but rather followed natural laws.

**Newton's Laws of Motion:** Sir Isaac Newton's laws of motion and universal gravitation, formulated in the late 17th century, provided a robust framework for understanding the behavior of celestial bodies. These laws explained the dynamics of planetary motion and demonstrated that planets moved in predictable orbits governed by gravity.

**Laplace's Nebular Hypothesis:** In the late 18th century, Pierre-Simon Laplace's nebular hypothesis gained prominence as a compelling explanation for the formation of planetary systems, including our solar system. This hypothesis replaced the notion of spontaneous generation with a systematic process of planetary formation from a spinning disk of gas and dust.

### Establishment of Modern Planetary Formation Theories

By the 19th century, the theory of spontaneous generation

of planets had lost its scientific credibility. The emergence of modern planetary formation theories, rooted in observational evidence and the laws of physics, solidified our understanding of how planets form. Notable developments in this transition included:

**Advancements in Telescopes:** Technological advancements in telescopes allowed astronomers to observe planets and other celestial objects with increasing detail, leading to a deeper understanding of their properties and origins.

**Solar System Exploration:** Space missions, such as those to study the Moon and other planets, provided direct evidence of planetary features and geological processes that were inconsistent with the idea of spontaneous generation.

**Planetary Science:** The establishment of planetary science as a distinct field of study led to comprehensive research on planetary formation processes, including accretion from protoplanetary disks and the role of gravity in shaping planetary systems.

In conclusion, the theory of spontaneous generation of planets, once rooted in ancient beliefs and mystical thinking, gradually lost favor as scientific understanding advanced. Observations, the development of scientific laws, and the rise of modern planetary formation theories contributed to its disproof. Today, we have a comprehensive understanding of the systematic processes by which planets and celestial bodies form, grounded in empirical evidence and the laws of physics. The theory of spontaneous generation of planets remains a historical curiosity, reflecting the evolution of scientific thought over centuries.

# The Static Universe Model

The static universe model, also known as the steady-state theory, was a cosmological hypothesis that posited the existence of an unchanging, eternal universe. This theory proposed that matter was continuously created to maintain a constant density in the expanding universe. It stood in contrast to the Big Bang theory, which eventually gained widespread acceptance as the prevailing explanation for the origin and evolution of the universe. The static universe model, while once considered by some scientists, eventually faced significant challenges and was largely discredited. In this exploration, we will delve into the essence of the static universe model, its historical origins, its early proponents, the influential individuals who believed in its validity, the process by which it was ultimately disproved, and the timeline of its discrediting.

**The Essence of the Static Universe Model**

The static universe model, also known as the steady-state theory, proposed that the universe was unchanging and eternal. It suggested that the universe maintained a constant density as it expanded, with new matter continuously being created to fill in the gaps left by cosmic expansion. This theory stood in contrast to the Big Bang theory, which posited a finite age for the universe and an initial singularity from which it expanded.

## Origins of the Static Universe Model

The origins of the static universe model can be traced back to the mid-20th century, a period marked by significant developments in cosmology.

**Hermann Bondi, Thomas Gold, and Fred Hoyle:** The static universe model was developed primarily by three British astrophysicists: Hermann Bondi, Thomas Gold, and Fred Hoyle. In 1948, they presented their theory in a series of papers collectively known as the "B2FH" paper, named after their initials.

## Proponents of the Static Universe Model

The static universe model had influential proponents within the scientific community during the mid-20th century.

**Fred Hoyle:** Fred Hoyle was one of the key proponents of the steady-state theory. He was a highly respected astrophysicist known for his work on nucleosynthesis in stars and coined the term "Big Bang" in a somewhat derogatory manner, although he disagreed with the theory.

## Challenges and Discrediting of the Static Universe Model

The static universe model began to face challenges and criticism as new evidence and observations emerged, particularly in the field of cosmology.

**Cosmic Microwave Background (CMB):** One of the most significant pieces of evidence against the static universe model came in the form of the discovery of the cosmic microwave background radiation. In the 1960s, astronomers Arno Penzias and Robert Wilson detected this faint, uniform radiation, which is a remnant of the early hot universe predicted by the Big Bang theory. The existence of the CMB provided strong support for the Big Bang model and contradicted the predictions of the steady-state theory.

**Expanding Universe:** The observation that galaxies were receding from each other, as evidenced by the redshift of their light, further supported the idea of an expanding universe with a finite origin. This observation was consistent with the predictions of the Big Bang theory but challenged the static universe model's assumption of a constant universe.

**Degradation of Support:** As the evidence against the static universe model accumulated, support for the theory among scientists waned. The model became increasingly less accepted within the scientific community.

### When the Static Universe Model Was Disproved

The static universe model gradually lost favor and was largely discredited during the 1960s and 1970s.

**1960s:** The discovery of the cosmic microwave background radiation and the compelling evidence of an expanding universe, along with the lack of observational support for continuous matter creation, led to the decline of the static universe model.

**1970s:** By this decade, the Big Bang theory had gained overwhelming support, and the steady-state theory was no longer considered a viable explanation for the nature and origin of the universe.

### Legacy and Insights

The static universe model, although ultimately disproved, played a role in the evolution of cosmology by providing an alternative hypothesis to the expanding universe and the Big Bang theory. It highlights the importance of empirical evidence and the dynamic nature of scientific progress. The steady-state theory's gradual decline and replacement with the Big Bang theory exemplify the scientific method in action—hypotheses are proposed, tested against observations, and either supported or refuted based on empirical evidence. The legacy of the static universe model lies in its contribution to the ongoing quest to understand the cosmos and the universe's true nature. It serves as a reminder that scientific theories, no matter how influential their proponents, are subject to revision in the light of new evidence and improved understanding.

# The Tychonian System

The Tychonian system, also known as the Tychonic model, was a geocentric cosmological theory that emerged during a time of profound scientific and philosophical change in the 16th century.  Named after the Danish astronomer Tycho Brahe, this theory represented a compromise between the heliocentric model proposed by Copernicus and the traditional geocentric view.  In the Tychonian system, the Earth remained at the center of the universe, but other planets revolved around the Sun.  Although it may seem like an attempt to bridge the gap between old and new ideas, the Tychonian system eventually gave way to the heliocentric model, which more accurately explained the observed movements of celestial bodies. In this exploration, we will delve into the essence of the Tychonian system, its historical origins, its proponents, the influential individuals who believed in its validity, the process by which it was ultimately disproved, and the timeline of its decline.

**The Essence of the Tychonian System**

The Tychonian system was a geocentric cosmological model that sought to reconcile the centuries-old Ptolemaic view, where the Earth was the center of the universe, with the emerging heliocentric model, where the Sun occupied the central position. In this compromise theory:

The Earth was positioned at the center of the universe, as in the geocentric model.

The Sun orbited the Earth, along with the Moon and other celestial bodies.

However, other planets, such as Mars, Jupiter, and Saturn, were proposed to revolve around the Sun.

This arrangement allowed the Tychonian model to explain the observed phenomena, such as the retrograde motion of planets, while still accommodating the geocentric belief that Earth held a privileged position in the cosmos.

## Origins of the Tychonian System

The Tychonian system originated in the late 16th century, a time marked by significant advancements in astronomy and the clash between traditional and emerging cosmological ideas.

**Tycho Brahe:** The Tychonian system is named after the Danish astronomer Tycho Brahe, who was a prolific observer of celestial phenomena. Tycho's precise and detailed observations of planetary positions paved the way for this new cosmological model.

## Proponents of the Tychonian System

The Tychonian system had influential proponents during its brief period of consideration.

**Tycho Brahe:** As the creator of the model, Tycho Brahe was its most prominent advocate. His meticulous observations of celestial bodies were central to the development of the Tychonian system.

**Johannes Kepler:** Although Kepler is primarily known for his laws of planetary motion that supported the heliocentric model, he initially considered the Tychonian system as a potential explanation for celestial motions. However, he eventually abandoned it in favor of the heliocentric model, leading to one of the most significant breakthroughs in astronomy.

**Discrediting of the Tychonian System**

The Tychonian system gradually lost favor and was replaced by the heliocentric model.

**Johannes Kepler's Laws:** Kepler's first law, which described the elliptical orbits of planets around the Sun, provided a more accurate explanation of planetary motion than the Tychonian system. This marked a significant departure from the geocentric perspective.

**Galileo's Observations:** Galileo Galilei's telescopic observations of Jupiter's moons and the phases of Venus provided further evidence in support of the heliocentric model, as they directly contradicted the Tychonian system.

**Development of the Heliocentric Model:** While the Tychonian system was an attempt to bridge the gap between geocentrism and heliocentrism, it could not compete with the mathematical elegance and explanatory power of the heliocentric model proposed by Copernicus and later refined by Kepler and Galileo.

**When the Tychonian System Was Disproved**

The Tychonian system gradually lost support during the late 16th and early 17th centuries.

**Early 17th Century:** The heliocentric model gained increasing acceptance among astronomers and scientists. Tycho Brahe's own observations, which he had left to Johannes Kepler, played a pivotal role in the eventual acceptance of the heliocentric view.

**Legacy and Insights**

The Tychonian system, while ultimately disproved, played a transitional role in the history of astronomy. It marked a moment when science was grappling with fundamental changes in our understanding of the cosmos. The shift from geocentrism to heliocentrism challenged deeply entrenched beliefs about humanity's place in the universe. The Tychonian system serves as a reminder of the complexities of scientific revolutions and the importance of empirical evidence in shaping our understanding of the natural world. Tycho Brahe's meticulous observations, though initially used to support the Tychonian system, ultimately contributed to the overthrow of geocentrism and the triumph of the heliocentric model. In this historical context, the Tychonian system underscores the dynamic nature of scientific progress and the power of evidence-based inquiry to transform our understanding of the cosmos.

# Flat Earth Theory

Flat Earth theory, also known as the flat Earth hypothesis, is a belief that contradicts the conventional understanding of our planet's shape. It postulates that the Earth is not a spherical object but rather a flat, disc-like structure. Although this theory has long been debunked and contradicted by a wealth of scientific evidence, it continues to garner attention from various groups and individuals. In this exploration, we will delve into the essence of Flat Earth theory, its historical origins, its early proponents, the individuals who have supported it, the process by which it was eventually disproved, and the timeline of its discrediting.

## The Essence of Flat Earth Theory

Flat Earth theory proposes that the Earth is not a spherical object, as modern science has demonstrated, but rather a flat, disc-like structure. According to this belief, the Earth's surface extends infinitely in all directions and is not curved. This theory contradicts the well-established understanding of Earth as an oblate spheroid.

## Origins of Flat Earth Belief

The notion of a flat Earth has ancient roots and can be traced back to early civilizations. Many early cultures, including the ancient Egyptians and Mesopotamians, depicted the Earth as flat in their cosmological beliefs. However, it wasn't until the ancient Greeks that the idea of a spherical Earth gained prominence.

**Early Notions of a Flat Earth:** Some early beliefs in a flat Earth can be attributed to limited observational capabilities. From the perspective of individuals on the Earth's surface, the terrain appeared flat, and there was no immediate visual evidence to suggest otherwise.

**The Greek Shift to Spherical Earth:** Ancient Greek philosophers, notably Pythagoras and later Parmenides, began to posit the idea of a spherical Earth based on observations such as the Earth's round shadow during lunar eclipses. The concept was further developed by philosophers like Plato and Aristotle, who provided both observational and theoretical arguments for a spherical Earth.

**Proponents of Flat Earth Belief**

While the spherical Earth model gained prominence in the ancient world, there were still proponents of flat Earth belief in various cultures throughout history. These proponents often relied on limited observations and cosmological beliefs rather than scientific evidence.

**Medieval and Renaissance Flat Earth Believers:** During the Middle Ages and Renaissance, there were individuals who continued to advocate for a flat Earth. Some of these beliefs were influenced by religious or philosophical considerations rather than empirical science.

**The Disproof of Flat Earth Theory**

Flat Earth theory began to unravel as scientific knowledge advanced, and more evidence supporting the spherical Earth model emerged. Several key developments contributed to the disproof of this theory.

**Circumnavigation:** One of the earliest pieces of evidence supporting the spherical Earth was the ability to circumnavigate the globe. As explorers like Ferdinand Magellan and subsequent voyagers sailed around the world, they observed that they returned to their starting point, which would not be possible on a flat Earth.

**Observations of the Horizon:** The ability to see distant objects gradually disappear below the horizon as a ship sails away provided further evidence for a curved Earth. This phenomenon would not occur on a flat surface.

**Curvature of Star Paths:** Observations of the night sky also contradicted a flat Earth. The positions of stars and constellations in the sky change depending on the observer's location on Earth, which is consistent with a spherical Earth but not with a flat one.

**Space Exploration:** The advent of space exploration in the mid-20th century provided irrefutable evidence of Earth's spherical shape. Photographs of Earth from space, taken by astronauts and satellites, clearly show our planet as a round object.

**When Flat Earth Theory Was Disproved**

While there were signs of the Earth's spherical shape dating back to ancient Greece, the definitive disproof of flat Earth theory occurred over a span of centuries, culminating in the 20th century with space exploration.

**Columbus' Voyages (Late 15th Century):** Christopher Columbus's voyages to the Americas in the late 15th century provided strong evidence for a spherical Earth. He set sail westward and eventually reached the Americas, demonstrating that the Earth was not an infinite expanse but rather a round planet.

**Space Age (20th Century):** The 20th-century space age, marked by the launch of artificial satellites and human spaceflight, provided unequivocal visual evidence of Earth's spherical shape. Astronauts who orbited the Earth and later traveled to the Moon captured photographs that definitively depicted a round planet.

**Legacy and Insights**

The discrediting of flat Earth theory highlights the importance of empirical evidence, observation, and the scientific method in understanding the nature of our planet. While this theory has persisted among certain groups, it is firmly rejected by mainstream science. The acceptance of a spherical Earth has been crucial in advancing our understanding of astronomy, navigation, and the natural world. Flat Earth theory remains a historical curiosity and a reminder of the enduring power of scientific inquiry and evidence-based thinking.

# Geocentrism

In the realms of cosmology, few ideas have held as much sway and sparked as much debate as geocentrism. This age-old theory, rooted in the notion that the Earth occupies the central position in the cosmos, shaped humanity's understanding of the universe for centuries. As we embark on this historical journey, we will delve into what geocentrism entails, its origins in antiquity, the minds behind its formulation, the influential figures who endorsed it, the eventual unraveling of this belief, and the momentous era when it was ultimately disproven.

## The Heart of Geocentrism: What Is It?

At its core, geocentrism is the cosmological concept that places the Earth at the center of the universe. According to this view, all celestial bodies, including the Sun, the Moon, and the stars, orbit around our planet. It's a theory that lends an undeniable sense of cosmic importance to Earth itself, casting it as the focal point of creation.

## Antiquity's Gaze: The Origins of Geocentrism

The roots of geocentrism stretch far back into the annals of antiquity. This ancient worldview can be traced to early civilizations, but it was the ancient Greeks who laid the intellectual foundation for geocentrism. The likes of Aristotle and Ptolemy, revered scholars of their time, contributed significantly to this perspective, crafting detailed models and theories that reinforced the notion of an Earth-centered cosmos.

## The Luminary Minds: Pioneers of Geocentrism

In the annals of scientific history, we encounter luminaries who embraced and championed the geocentric model. Among these was Claudius Ptolemy, a Greco-Roman mathematician, astronomer, and geographer who authored the influential work "Almagest" in the 2nd century. Ptolemy's geocentric model of the cosmos, known as the Ptolemaic system, became the prevailing view for over a thousand years. The geocentric theory found resonance in the works of philosophers like Aristotle and scholars throughout the medieval period.

## The Celestial Revolution: The Unraveling of Geocentrism

But as history unfurled, the geocentric model began to face increasing scrutiny. It was a time when curious minds dared to challenge established beliefs. One such individual was Nicolaus Copernicus, a Renaissance mathematician and astronomer. In the early 16th century, Copernicus proposed a radical idea: a heliocentric model in which the Sun, not the Earth, was at the center of the cosmos. His work "De revolutionibus orbium coelestium" (On the Revolutions of the Celestial Spheres), published in 1543, marked a turning point in the history of cosmology.

## The Heliocentric Revelation: The End of an Era

The heliocentric model presented by Copernicus triggered a profound shift in our understanding of the cosmos. However, it wasn't until the advent of the telescope and the meticulous observations of astronomers like Johannes Kepler and Galileo

Galilei that the heliocentric view gained significant ground.

Kepler's laws of planetary motion, developed in the early 17th century, provided mathematical support for the heliocentric model. Galileo's telescopic observations revealed evidence such as the phases of Venus and the moons of Jupiter, which contradicted the geocentric view. These groundbreaking findings challenged the entrenched geocentric paradigm.

## A New Dawn: The Disproval of Geocentrism

The ultimate disproof of geocentrism came when Sir Isaac Newton formulated his laws of motion and universal gravitation in the late 17th century. These laws not only provided a robust framework for understanding celestial motion but also offered compelling evidence for the heliocentric model. The gravitational forces exerted by the Sun on planets and other celestial bodies could be described mathematically, further solidifying the heliocentric view.

In essence, the geocentric model, which had dominated scientific and philosophical thought for centuries, gave way to the heliocentric understanding of the cosmos. The geocentric theory, with its Earth-centered perspective, was officially debunked by the mid-17th century.

## A Cosmic Shift: The Legacy of Geocentrism

The transition from geocentrism to heliocentrism marked a profound moment in the history of science. It challenged deeply ingrained beliefs and reshaped humanity's perception of its place in the universe. The heliocentric model, with the Sun at its center, became the foundation of modern astronomy and our understanding of celestial mechanics.

In conclusion, geocentrism, an ancient cosmological concept positioning the Earth at the center of the universe, had its origins in antiquity and was championed by renowned figures like Claudius Ptolemy. However, the heliocentric model proposed by Copernicus, supported by Kepler, Galileo, and ultimately Newton, led to the eventual disproof of geocentrism in the 17th century. This transition marked a pivotal moment in the history of science, altering humanity's cosmic perspective and paving the way for modern astronomy.

# Catastrophic Plate Tectonics Theory

The Catastrophic Plate Tectonics Theory was a scientific hypothesis aimed at explaining the rapid movement of Earth's tectonic plates and major geological events. It proposed that tectonic plate movements were once much more rapid, leading to catastrophic events like the biblical Flood. This theory, though once entertained by a small group of scientists, ultimately succumbed to a lack of supporting evidence and inconsistencies with established geological principles. In this exploration, we will delve into what the theory entailed, its origins, the key proponents, its gradual decline in credibility, and when it was effectively disproven.

**What Was the Catastrophic Plate Tectonics Theory?**

The Catastrophic Plate Tectonics Theory was an attempt to explain how the Earth's continents and tectonic plates moved in the past. According to this hypothesis, the present-day slow drift of continents was preceded by a period of rapid, catastrophic plate movements. This rapid plate tectonics was believed to have been responsible for numerous geological phenomena, including the formation of mountain ranges, ocean basins, and the rapid shifting of continents.

One of the most notable aspects of this theory was its connection to the biblical Flood. Some proponents of Catastrophic Plate Tectonics suggested that this rapid plate movement was the cause of the great deluge described in religious texts.

### Origins and Key Proponents

The Catastrophic Plate Tectonics Theory emerged in the late 20th century as an alternative to the conventional theory of plate tectonics. While it was not the work of a single individual, several scientists played crucial roles in its development and promotion:

**John Baumgardner:** A geophysicist and computer modeler, Baumgardner was a prominent proponent of Catastrophic Plate Tectonics. He developed computer simulations to support the theory, suggesting that rapid plate movements could explain geological features.

**Steven Austin:** A geologist and creationist, Austin was another advocate of the theory. He proposed that the rapid movement of tectonic plates during the Flood could account for various geological formations.

### Scientific Reception and Challenges

The Catastrophic Plate Tectonics Theory faced significant challenges from the scientific community and several inherent problems:

**Lack of Mechanism:** Critics argued that the theory lacked a plausible mechanism to explain the sudden acceleration of plate movements. The energy required for such rapid shifts would have been immense, and the source of this energy remained unclear.

**Heat Generation:** Rapid plate movements would have generated enormous amounts of heat, enough to melt the Earth's crust. The lack of evidence for widespread melting and the presence of ancient rocks inconsistent with extreme heating posed substantial problems for the theory.

**Fossil Record:** The theory struggled to account for the preservation of delicate fossils and sedimentary layers during rapid plate movements. Such movements should have destroyed evidence of ancient life and disrupted sedimentary sequences.

**Geological Evidence:** Many geological features and observations, such as the slow formation of mountain ranges and the gradual spreading of ocean basins, were inconsistent with the rapid plate tectonics proposed by the theory.

### Decline and Disproof

The Catastrophic Plate Tectonics Theory gradually lost credibility within the scientific community due to these insurmountable challenges and a lack of supporting evidence. As more research was conducted in the fields of geology, geophysics, and paleontology, the conventional theory of plate tectonics continued to provide a more comprehensive and consistent explanation for Earth's geological processes.

While it is difficult to pinpoint an exact moment when the theory was definitively disproven, it became increasingly marginalized as it failed to gain acceptance among mainstream scientists. The scientific community overwhelmingly supported the conventional plate tectonics theory, which is backed by a wealth of empirical evidence and continues to be refined through ongoing research.

Today, the Catastrophic Plate Tectonics Theory is primarily associated with certain creationist and pseudoscientific circles. It serves as a historical example of an alternative scientific hypothesis that ultimately could not withstand the scrutiny of the broader scientific community. As with many fringe theories, its proponents continue to advocate for it, but it remains on the outskirts of mainstream geological science.

In summary, the Catastrophic Plate Tectonics Theory was an alternative hypothesis aimed at explaining rapid plate movements and geological events, often associated with the biblical Flood. Despite initial interest from a small group of scientists, it encountered insurmountable challenges, lacked supporting evidence, and gradually declined in credibility. While it cannot be pinpointed to a specific moment, the theory faded from mainstream scientific consideration as conventional plate tectonics continued to provide a more robust and consistent explanation for Earth's geological processes.

# The Hollow Moon Theory

The Hollow Moon Theory, a captivating but debunked hypothesis, suggests that Earth's moon is not a solid celestial body but rather a hollow, artificial structure. While this theory has intrigued some for centuries, it lacks scientific validity and stands as a testament to the enduring allure of unconventional ideas. In this exploration, we will delve into the origins of the Hollow Moon Theory, its proponents, the eventual disproof of its claims, and its place in the annals of pseudoscience.

**Origins of the Hollow Moon Theory**

The Hollow Moon Theory posits that the moon is not a naturally formed celestial object but rather an artificial construct, either crafted by advanced extraterrestrial beings or created by humans in the distant past. It suggests that the moon's hollow interior houses advanced technologies or serves as a base for extraterrestrial civilizations.

**Early Speculations and Sci-Fi Influence**

The origins of the Hollow Moon Theory can be traced back to early human fascination with the moon and speculative fiction. Throughout history, various cultures have woven myths and legends around the moon, often attributing mystical properties to it. While these myths do not directly relate to the modern Hollow Moon Theory, they underscore humanity's enduring fascination with Earth's celestial neighbor.

The rise of science fiction literature in the 19th and 20th centuries further fueled ideas about artificial celestial bodies, including the moon. Works like H.G. Wells' "The First Men in the Moon" (1901) and Jules Verne's "From the Earth to the Moon" (1865) explored fantastical ideas of moon colonization and exotic lunar landscapes.

**Early Proponents**

While the Hollow Moon Theory has no single originator, it gained traction in the mid-20th century among certain fringe groups and pseudoscientists. These individuals proposed various scenarios to support their claims:

**Nikolai A. Kozyrev (1908-1983):** A Soviet astronomer, Kozyrev suggested that the moon's craters were artificial constructs created by an ancient, technologically advanced civilization. He based this hypothesis on his interpretation of lunar photographs.

**Michael Vasin and Alexander Shcherbakov:** In the 1970s, these Soviet scientists speculated that the moon was a hollow spacecraft constructed by extraterrestrial beings. They proposed that the moon's craters were the result of its propulsion system.

**George H. Leonard (1929-2010):** Leonard, an engineer, authored the book "Somebody Else is on the Moon" in 1976. He claimed that NASA was covering up evidence of alien structures on the moon and that it was artificially constructed.

**Disproof and Scientific Rebuttals**

The Hollow Moon Theory lacks scientific validity and has been widely debunked by the scientific community. The following points represent some of the key scientific arguments against this theory:

**Gravitational Effects:** The moon's gravitational effects on Earth, such as tides, would not be possible if it were hollow. The moon's mass and gravitational pull are consistent with it being a solid celestial body.

**Lunar Seismology:** Data from seismometers placed on the moon during the Apollo missions provided valuable insights into its internal structure. These measurements indicated that the moon has a solid, layered composition, not a hollow interior.

**Lunar Samples:** The rock and soil samples brought back from the moon by Apollo astronauts provide physical evidence of a solid lunar surface.

**Orbital Dynamics:** The moon's orbit and interactions with other celestial bodies can be accurately modeled based on its mass and density, further confirming its solid nature.

**Lunar Crater Formation:** The moon's craters are the result of meteoroid impacts over billions of years, a process well-understood in planetary science.

**Modern Understanding and Legacy**

While the Hollow Moon Theory has been thoroughly discredited, it endures as an example of pseudoscientific thinking. Its continued existence within fringe circles demonstrates the persistent appeal of unconventional and often fantastical ideas. In the face of overwhelming scientific evidence, the theory remains a relic of a bygone era when speculation often outpaced empirical investigation.

In conclusion, the Hollow Moon Theory, which posits that the moon is a hollow, artificial construct, has no basis in scientific reality. It emerged from a blend of ancient myths, speculative fiction, and pseudoscientific ideas in the 20th century. Despite gaining traction among certain groups, it has been debunked through empirical evidence, such as lunar seismology and the analysis of lunar samples. The Hollow Moon Theory serves as a cautionary example of how pseudoscientific beliefs can capture the imagination but ultimately fall short in the face of scientific scrutiny.

# The Hollow Earth Theory

The Hollow Earth theory is one of the most intriguing and enduring pseudoscientific ideas in the history of human thought. This concept proposes that the Earth is not a solid sphere but rather a hollow shell with openings at its poles. Within this hollow Earth, it is believed by some that there might exist subterranean realms, hidden civilizations, and unique ecosystems. The Hollow Earth theory has fascinated explorers, writers, and thinkers for centuries, giving rise to a range of imaginative tales and captivating narratives. In this exploration, we will delve into the essence of the Hollow Earth theory, its historical origins, the early proponents, the notable figures who embraced this concept, the process by which it was eventually discredited, and the timeline of its decline.

## The Essence of the Hollow Earth Theory

The Hollow Earth theory posits that our planet, instead of being a solid mass of rock and molten metal, is hollow with a vast interior space. The interior is believed to be illuminated by a central sun, providing light and heat to a hidden world beneath the Earth's surface. This theory often suggests the existence of polar openings, through which one could access the inner Earth.

## Origins of the Hollow Earth Theory

The roots of the Hollow Earth theory can be traced back to various ancient and medieval legends, but it gained significant attention during the Enlightenment and the subsequent centuries.

**Edmond Halley:** The theory is often associated with the English scientist Edmond Halley, best known for calculating the orbit of Halley's Comet. In the late 17th century, Halley proposed that the Earth might consist of multiple concentric shells, each with its own atmosphere and magnetic poles.

## Proponents of the Hollow Earth Theory

Over the centuries, numerous individuals and groups embraced the Hollow Earth theory and contributed to its development.

**John Cleves Symmes Jr.:** In the early 19th century, the American army officer and lecturer John Cleves Symmes Jr. gained notoriety for his passionate advocacy of the Hollow Earth theory. Symmes claimed that the Earth had openings at its poles and that these openings could be accessed by daring explorers.

**Cyrus Teed:** In the late 19th and early 20th centuries, Cyrus Teed, also known as Koresh, developed a variation of the Hollow Earth theory known as Koreshanity. He believed that we lived inside a concave Earth with the celestial bodies and stars located within the Earth's shell.

**Admiral Richard E. Byrd:** Although primarily known as a pioneering aviator and polar explorer, Admiral Richard E. Byrd's expeditions to Antarctica in the 20th century fueled speculation about Hollow Earth. Some theorists claim that Byrd's expeditions were an attempt to enter the inner Earth through polar openings.

**Discrediting of the Hollow Earth Theory**

The Hollow Earth theory has been largely discredited through scientific observations and a deeper understanding of Earth's composition.

**Geological Knowledge:** As geology advanced, our understanding of Earth's interior structure became more precise. Seismic studies, gravity measurements, and the behavior of seismic waves during earthquakes all provided evidence against a hollow Earth.

**Modern Physics:** Modern physics and astronomy have further confirmed our understanding of planetary formation, and the Hollow Earth theory is inconsistent with these established principles.

**Space Exploration:** The exploration of outer space and the study of celestial bodies have provided insights into the fundamental nature of planets, including the Earth. The Hollow Earth theory does not align with the knowledge gained from space missions.

**When the Hollow Earth Theory Was Disproved**

The gradual discrediting of the Hollow Earth theory took place over several centuries, with key developments occurring in the 19th and 20th centuries.

**19th Century:** As scientific knowledge expanded during the 19th century, the Hollow Earth theory began to lose credibility among mainstream scientists.

**20th Century:** With advancements in geology, physics, and space exploration, the Hollow Earth theory became increasingly untenable. By the mid-20th century, it was largely relegated to the realm of pseudoscience and science fiction.

**Legacy and Insights**

The Hollow Earth theory remains a fascinating example of human imagination and the enduring allure of the unknown. While it has been largely discredited by scientific knowledge, it continues to capture the imagination of some individuals and persists in various forms in literature and popular culture. The story of the Hollow Earth theory reminds us of the human capacity for exploration, curiosity, and the enduring appeal of mysteries hidden beneath the surface of our world. It also underscores the importance of empirical evidence, critical thinking, and the scientific method in discerning fact from fiction in the quest for understanding the universe around us.

# The Steady State Universe

**Understanding the Steady State Universe**

The Steady State Universe was a cosmological theory that proposed an unchanging and eternal universe. Unlike the Big Bang theory, which suggests a universe with a definite beginning in a cosmic explosion, the Steady State Universe posited that the universe had no beginning or end. It maintained a constant density over time through a process of continuous creation of matter.

**Origins and Early Development**

The idea of a Steady State Universe emerged in the mid-20th century, primarily in response to the growing evidence supporting the expanding universe and the Big Bang theory. The theory's development can be attributed to the collaboration of several prominent scientists:

**Hermann Bondi:** A British-Austrian mathematician and cosmologist, Bondi was one of the key proponents of the Steady State theory. He, along with colleagues Thomas Gold and Fred Hoyle, formulated the theory as an alternative to the Big Bang model.

**Thomas Gold:** An Austrian-American astrophysicist, Gold played a significant role in developing the Steady State Universe theory alongside Bondi and Hoyle. He contributed to the theory's mathematical formulation.

**Fred Hoyle:** A British astronomer and cosmologist, Hoyle was another influential figure in the development of the Steady State theory. He coined the term "Big Bang" during a BBC radio broadcast in a somewhat dismissive manner.

### Believers and Supporters

The Steady State Universe theory gained some notable adherents in the scientific community during the mid-20th century. This support was largely in response to the desire for an alternative to the Big Bang theory. Prominent scientists and astronomers who supported or explored the Steady State theory included:

**Fred Hoyle:** Hoyle was not only a proponent but also a fervent advocate of the Steady State Universe throughout his career.

**Hermann Bondi:** As one of the theory's co-developers, Bondi was a vocal supporter of the Steady State model.

**Thomas Gold:** Gold contributed to the development of the theory and supported its ideas.

### The Story of How the Theory Eventually Got Disproved

The Steady State Universe theory faced several challenges and ultimately succumbed to mounting evidence from astrophysical observations and theoretical developments:

**Cosmic Microwave Background (CMB):** One of the most compelling pieces of evidence against the Steady State theory came from the discovery of the cosmic microwave background radiation. In 1965, Arno Penzias and Robert Wilson stumbled upon this faint radiation, which was predicted by the Big Bang theory but not expected in a Steady State Universe. The CMB provided strong support for the idea that the universe had a hot and dense beginning.

**Observational Evidence:** Astronomical observations, including the redshift of distant galaxies and the abundance of light elements, consistently favored the expanding universe model proposed by the Big Bang theory.

**Theoretical Challenges:** As cosmological models and our understanding of the universe evolved, the Steady State theory struggled to account for new observations and evidence. It faced increasing difficulties in explaining the observed changes in the universe's structure and composition.

**When the Steady State Universe Theory Was Disproved**

The decline of the Steady State Universe theory occurred gradually but can be marked by significant milestones:

**1965:** The discovery of the cosmic microwave background radiation by Penzias and Wilson provided compelling evidence against the Steady State theory.

**Late 1960s to 1970s:** Astronomical observations and theoretical developments continued to support the expanding universe model of the Big Bang theory.

**Late 20th Century:** By the late 20th century, the Steady State theory had lost favor among most cosmologists and scientists. The overwhelming consensus in the scientific community shifted toward the Big Bang theory as the most accurate description of the universe's origin and evolution.

The Steady State Universe theory gradually faded from prominence in cosmology, replaced by the widely accepted Big Bang theory, which explained the universe's expansion and its origins in a hot and dense state approximately 13.8 billion years ago.

**Legacy and Insights**

While the Steady State Universe theory ultimately proved

to be incorrect, it played a valuable role in the history of cosmology. It encouraged scientific inquiry and debate, highlighting the importance of empirical evidence and the willingness to challenge prevailing theories in the pursuit of understanding the cosmos. The Steady State theory's rise and fall underscore the dynamic and evolving nature of scientific knowledge and our ongoing quest to unravel the mysteries of the universe.

# III

# Biology

# Spontaneous Human Combustion

Spontaneous Human Combustion (SHC) is a perplexing and controversial phenomenon that has captured the imaginations of many for centuries. In this exploration, we will delve into what SHC entails, its historical origins, the elusive identity of its originator, notable figures who have entertained its possibility, the gradual debunking of the theory, and when it began to lose its credibility.

**The Essence of Spontaneous Human Combustion**

Spontaneous Human Combustion refers to the unexplained and rare phenomenon in which a human body allegedly ignites and burns without any external source of ignition, such as an open flame or spark. Typically, only a small portion of the body is consumed by the fire, leaving surrounding objects relatively untouched. It is important to note that SHC is not accepted as a scientific phenomenon and is considered pseudoscientific due to a lack of empirical evidence.

**Origins of Spontaneous Human Combustion**

The origins of the concept of Spontaneous Human Combustion can be traced back several centuries. Reports of individuals mysteriously bursting into flames date back to the 17th century.

**Ambroise Paré:** While not the originator of the theory,

Ambroise Paré, a French surgeon and anatomist from the 16th century, described cases resembling SHC in his work. Paré attributed these incidents to the presence of an internal flammable substance, suggesting a natural cause rather than supernatural.

### The Elusive Originator

The concept of Spontaneous Human Combustion does not have a specific originator, and its development is attributed to the collective observations and reports of unusual deaths over time. It emerged as an explanation for inexplicable deaths marked by localized burns in the absence of external sources of ignition.

### Notable Figures and Believers

Throughout history, there have been instances where notable figures and individuals entertained the possibility of Spontaneous Human Combustion. While not widespread, some believed in its occurrence or at least considered it as a potential explanation for certain unexplained deaths.

**Charles Dickens:** The famous novelist Charles Dickens mentioned a case of alleged SHC in his novel "Bleak House," published in 1853. Dickens' description of the phenomenon in his novel may have contributed to its popularization.

### Debunking Spontaneous Human Combustion

Spontaneous Human Combustion has been met with skepticism and debunking by experts in various fields, including medicine, physics, and forensic science. The debunking of SHC primarily revolves around the following key arguments:

**No Scientific Evidence:** One of the most significant challenges to the SHC theory is the lack of scientific evidence supporting its existence. Claims of SHC are often based on anecdotal accounts rather than empirical data.

**Ignition Sources:** Skeptics argue that there must be an external ignition source for combustion to occur. Without a spark or open flame, it is unlikely for a human body to ignite spontaneously.

**Alternative Explanations:** Many cases attributed to SHC can be explained by more plausible factors, such as accidents involving candles, smoking, or faulty electrical appliances.

**Natural Processes:** Some experts propose that the combustion of a human body can be explained through natural processes, such as the "wick effect." In this scenario, a small external ignition source (like a cigarette) starts a fire, and the body's fat acts as a fuel source, sustaining the flames.

### Loss of Credibility

Over time, as scientific knowledge and forensic techniques have advanced, belief in Spontaneous Human Combustion has waned. The phenomenon began to lose credibility as alternative explanations for cases once attributed to SHC became more widely accepted.

**Forensic Science:** Advances in forensic science have allowed for more accurate investigations into unusual deaths. In cases that were previously labeled as SHC, forensic experts have been able to identify more rational causes for the burns, such as accidental fires or even homicide.

**Critical Analysis:** Critical analysis and scrutiny by the scientific community have contributed to the diminishing belief in SHC. Skeptical examination of the available evidence and the application of scientific principles have led to the rejection of the theory.

### Conclusion

Spontaneous Human Combustion, a centuries-old phenomenon characterized by unexplained fires igniting within

the human body, has remained a subject of fascination and intrigue. However, despite the persistent interest and occasional
notoriety of individual cases, the theory has lost credibility in
the eyes of the scientific community. The absence of empirical
evidence, the development of alternative explanations, and
advances in forensic science have collectively contributed
to the debunking of SHC. Today, it remains an example of
how scientific scrutiny and rational inquiry can dispel long-
standing mysteries and pseudoscientific beliefs.

# Animal Magnetism

Animal magnetism, a pseudoscientific theory, emerged in the late 18th century and was popularized by Franz Anton Mesmer, an Austrian physician. This theory proposed the existence of a magnetic fluid within living beings, which could be manipulated to treat various medical conditions. In this exploration, we will delve into what animal magnetism entails, its historical origins, Mesmer's role in its development, notable figures who believed in the theory, its eventual debunking, and the timeline of its decline.

## The Essence of Animal Magnetism

Animal magnetism is based on the concept of an invisible, natural force or fluid that flows through all living beings. This fluid was believed to influence a person's health and well-being. According to the theory, imbalances or blockages in this fluid could lead to illness, and the practitioner of animal magnetism, often referred to as a "magnetizer," could restore balance by manipulating this fluid.

## Origins of Animal Magnetism

Animal magnetism found its origins in the late 18th century, gaining prominence in Europe. The term "animal magnetism" itself was coined by Friedrich Anton Mesmer, who played a pivotal role in its development and popularization.

**Franz Anton Mesmer: The Face of Animal Magnetism**

Franz Anton Mesmer, born in 1734 in Austria, is the figure most closely associated with the theory of animal magnetism. Mesmer initially studied medicine but later developed his theory of animal magnetism, which he believed could treat a wide range of medical conditions.

**Believers in Animal Magnetism**

Animal magnetism found followers and believers during the late 18th and early 19th centuries. Several notable figures, including scientists and physicians, were intrigued by Mesmer's theories and conducted experiments to explore the concept. Some believed that animal magnetism held promise as a therapeutic practice.

**Marquis de Puységur:** A disciple of Mesmer, Marquis de Puységur, expanded on the theory and introduced the concept of "magnetic somnambulism," wherein patients entered a trance-like state during magnetization.

**The Decline and Debunking of Animal Magnetism**

Despite its popularity, animal magnetism faced increasing skepticism and criticism as the 18th century progressed into the 19th century. Several key factors contributed to the decline and eventual debunking of the theory:

**Scientific Scrutiny:** As scientific knowledge advanced, animal magnetism came under increasing scrutiny. Scientists and physicians began to question the existence of the magnetic fluid and the efficacy of magnetization.

**Placebo Effect:** It became evident that many of the purported cures attributed to animal magnetism were likely due to the placebo effect rather than the manipulation of a magnetic fluid. Patients who believed in the treatment often experienced improvements in their conditions.

**Magnetic Theories:** Advances in the understanding of magnetism in the natural world also cast doubt on animal magnetism. The magnetic fluid proposed by Mesmer did not align with the principles of magnetism established in physics.

**Mesmer's Decline:** Franz Anton Mesmer himself faced personal and professional setbacks. His unorthodox methods and claims led to criticism and legal action. He eventually retreated from public life, and animal magnetism lost its most prominent advocate.

### Timeline of Decline

The decline of animal magnetism occurred gradually over the course of the 19th century. Some key milestones in its decline include:

**Early 19th Century:** Mesmer's decline in influence and the growing skepticism among the scientific community contributed to a decrease in the practice of animal magnetism.

**1820s:** The rise of more established and evidence-based medical practices further marginalized animal magnetism.

**Late 19th Century:** By the end of the 19th century, animal magnetism had largely fallen out of favor as a legitimate medical practice. It was increasingly viewed as a pseudoscience.

### Conclusion

Animal magnetism, with its belief in an invisible magnetic fluid flowing through living beings, was a pseudoscientific theory that gained popularity in the late 18th century, largely due to the efforts of Franz Anton Mesmer. While it attracted some followers and believers, it faced increasing skepticism as scientific knowledge advanced. Ultimately, the theory of animal magnetism was debunked and declined in influence throughout the 19th century. Today, it serves as a historical example of a once-prominent belief that could not withstand scientific scrutiny and critical examination.

# Transmutation of Species

The doctrine of transmutation of species, a precursor to the theory of evolution, is a fascinating chapter in the history of science. In this discussion, we will explore the essence of this doctrine, its origins, the early proponents, the historical figures who believed in its validity, the process through which it was eventually disproved, and when it gave way to the modern theory of evolution by natural selection.

**The Essence of the Doctrine of Transmutation of Species**

The doctrine of transmutation of species, often referred to simply as "transmutation," posits that species change over time and can transform into other species. In essence, it suggests that the diversity of life on Earth is not fixed but rather dynamic, with species evolving from one another. This theory challenged the prevailing belief in the fixity of species, which held that each species was created separately and remained unchanged throughout time.

**Origins of the Doctrine**

The origins of the doctrine of transmutation of species can be traced back to ancient civilizations where early thinkers pondered the diversity of life. However, the doctrine gained more structured and scientific attention in the 18th and 19th centuries.

**Ancient Ideas:** The concept of species transformation was hinted at in the works of ancient philosophers like Empedocles and Anaximander, who proposed that life forms could change over time.

**Lamarck's Contribution:** The theory of transmutation gained notable attention with the work of Jean-Baptiste Lamarck, a French biologist and early proponent of the idea. In his 1809 book "Philosophie Zoologique," Lamarck proposed that species could evolve through the inheritance of acquired characteristics. He believed that organisms developed new traits during their lifetime and passed these traits on to their offspring.

### Early Proponents and Believers

During the 19th century, the doctrine of transmutation gained a significant following among scientists and intellectuals. Some key figures who supported this theory included:

**Erasmus Darwin:** Charles Darwin's grandfather, Erasmus Darwin, was an early proponent of transmutation. In his writings, he proposed the idea that species changed over time through competition for resources.

**Geoffroy Saint-Hilaire:** French naturalist Étienne Geoffroy Saint-Hilaire was another advocate of transmutation. He suggested that species could transform into other species, and he engaged in debates with proponents of the fixity of species.

**Robert Grant:** Scottish biologist Robert Grant, who was Charles Darwin's teacher, also supported the idea of species transformation. Grant's ideas influenced Darwin's thinking on evolution.

**Alfred Russel Wallace:** Although not a proponent of Lamarckian transmutation, Alfred Russel Wallace independently developed a theory of natural selection that closely paralleled Charles Darwin's work.

**The Disproof of Transmutation and Emergence of Evolution by Natural Selection**

The doctrine of transmutation of species began to lose ground as scientific understanding advanced. Several factors contributed to its eventual disproof:

**Lack of Mechanism:** One of the major challenges for transmutation was the absence of a well-defined mechanism to explain how species could change over time. Lamarck's idea of acquired characteristics was criticized for its lack of empirical evidence.

**Fossil Record:** The growing body of evidence from the fossil record revealed that species did not appear to change gradually and uniformly, as transmutation suggested. Instead, the fossil record showed a pattern of extinction and the appearance of new, distinct species.

**Charles Darwin's Theory:** Charles Darwin, influenced by the works of Malthus, Lyell, and others, developed the theory of evolution by natural selection. His groundbreaking book, "On the Origin of Species," published in 1859, provided a comprehensive and testable mechanism for the diversity of life. Natural selection, based on the differential survival and reproduction of individuals with advantageous traits, explained the emergence of new species.

**The Transition to Modern Evolutionary Theory**

With the publication of Charles Darwin's work, the doctrine of transmutation of species gave way to modern evolutionary theory based on natural selection. Darwin's theory provided a robust framework supported by extensive evidence from comparative anatomy, embryology, biogeography, and the fossil record.

## Conclusion

The doctrine of transmutation of species, while an important precursor to modern evolutionary theory, ultimately faced scientific challenges and limitations. It lacked a clear mechanism and was unable to account for the patterns observed in the fossil record. Charles Darwin's theory of evolution by natural selection, which addressed these shortcomings, marked a turning point in the history of biology. Today, the doctrine of transmutation remains a historical curiosity, reflecting the evolution of scientific thought and the gradual emergence of our current understanding of the diversity of life on Earth.

## *Breatharianism Belief: A Controversial Ideology*

Breatharianism, also known as inedia, is a controversial belief that suggests individuals can live without consuming food and sometimes even water, relying solely on prana or life energy. In this exploration, we will delve into the essence of this theory, its origins, the individuals attributed to its development, any notable figures who believed in its validity, the eventual debunking of the theory, and when it began to lose credibility.

## The Essence of Breatharianism

Breatharianism posits that humans can subsist without traditional nourishment, such as food and water, by harnessing a form of spiritual or cosmic energy called prana. This concept stands in stark contrast to the well-established understanding of human biology, which dictates that sustenance is derived from the intake of calories, nutrients, and hydration.

## Origins of Breatharianism

The origins of breatharianism can be challenging to pinpoint precisely, as similar ideas have appeared in various cultures and historical contexts. However, the modern form of breatharianism as we know it began to gain attention in the late 20th century.

**Early Influences:** Elements of breatharianism can be traced back to ancient spiritual practices and the idea of fasting as a means of spiritual purification. Various religious traditions have practiced fasting as a form of discipline and purification for centuries.

**Jasmuheen:** One of the most prominent figures associated with modern breatharianism is Ellen Greve, who adopted the pseudonym Jasmuheen. She is often credited with popularizing the belief system in the late 20th century. Jasmuheen claimed that she could survive on prana alone, without the need for conventional sustenance.

## Notable Believers and Influencers

Breatharianism, despite its lack of scientific support, has attracted a small but dedicated following. Some individuals, often in pursuit of spiritual enlightenment or physical purity, have embraced this belief system. Notable figures who have advocated for or practiced breatharianism to varying degrees include:

**Jasmuheen:** As mentioned earlier, Jasmuheen is one of the most well-known breatharians. She authored several books on the subject and conducted workshops and seminars promoting her ideas.

**Wiley Brooks:** Wiley Brooks is another prominent figure in the breatharian movement. He founded the "Breatharian Institute of America" and claimed to subsist solely on sunlight and air.

**Hira Ratan Manek:** Also known as HRM, Manek claimed to live on solar energy and became known as the "Sun Gazer." He advocated gazing at the sun for nourishment.

## The Debunking of Breatharianism

Breatharianism has been widely criticized and debunked by medical professionals, scientists, and skeptics. The disproof of breatharianism is rooted in the fundamental principles of biology and physiology:

**Lack of Scientific Evidence:** Despite claims made by proponents, there is no scientific evidence to support the notion that humans can survive without food and water solely through the intake of prana or cosmic energy.

**Medical Dangers:** Extended periods without food and water can lead to severe health consequences, including malnutrition, organ failure, and death. These risks are well-documented in medical literature.

**Exposure of Fraud:** Some individuals who claimed to be breatharians have been exposed as frauds when they were observed consuming food and water in secret. These incidents have further eroded the credibility of the belief system.

## The Waning Credibility of Breatharianism

Over time, the belief in breatharianism has lost credibility in the eyes of the mainstream scientific and medical communities. The absence of empirical evidence, coupled with the documented dangers of prolonged fasting, has made it increasingly difficult for proponents to maintain their claims.

## Conclusion

Breatharianism, with its controversial premise of living solely on prana or cosmic energy without the need for traditional sustenance, has been a fringe belief that lacks scientific validity. While it has attracted a small number of followers over the years, the theory has faced widespread skepticism and debunking by experts in the fields of biology, medicine, and nutrition. Today, breatharianism remains on the periphery of accepted scientific understanding, a cautionary tale about the dangers of unfounded beliefs and practices.

# The Great Chain of Being

**Understanding the Great Chain of Being**

The Great Chain of Being is a philosophical and theological concept that was prevalent in Western thought for centuries. It was a way of understanding the hierarchy and order of the universe, from the lowest forms of existence to the highest, with God at the pinnacle. This concept encapsulates several key aspects:

**Hierarchy:** The Great Chain of Being depicts a hierarchical structure where every entity, from the tiniest speck of matter to the divine, occupies a specific place in the cosmic order. This hierarchy is fixed and unchanging.

**Order and Perfection:** Each level of the hierarchy was believed to represent a higher degree of order, complexity, and perfection. This implied that lower forms of existence were inherently less perfect than those higher up the chain.

**Immutability:** The Great Chain of Being was thought to be immutable, meaning that entities could not change their position within the hierarchy. This idea was deeply rooted in the philosophical and religious beliefs of the time.

**Origins and Early Development**

The origins of the Great Chain of Being can be traced back to ancient Greek philosophy, particularly the works of Plato and Aristotle. Plato's concept of the Forms or Ideas, which represented the ideal and unchanging essence of things, laid the groundwork for the idea of a hierarchical structure in the universe.

Aristotle, in his "Scala Naturae" or "Ladder of Life," introduced the idea of a natural hierarchy in the living world, with plants at the bottom, followed by animals, and ultimately humans. This hierarchy was based on the complexity of each group's soul.

### Prominent Figures and Believers

Throughout history, many prominent philosophers, theologians, and scholars embraced the Great Chain of Being as a central concept in their worldview. Notable figures include:

**Thomas Aquinas:** The medieval theologian Thomas Aquinas incorporated the Great Chain of Being into his synthesis of Christian theology and Aristotelian philosophy. He emphasized the unchanging order of the universe as evidence of God's divine plan.

**Renaissance Thinkers:** During the Renaissance, the concept of the Great Chain of Being was widely accepted. Thinkers like Giovanni Pico della Mirandola and Marsilio Ficino integrated it into their philosophical and mystical writings.

**Shakespearean Influence:** William Shakespeare's plays often reflect the idea of the Great Chain of Being. In "Hamlet," for example, the character Hamlet reflects on the hierarchical order of the universe.

### The Story of How the Theory Eventually Got Disproved

The decline of the Great Chain of Being as a dominant worldview was gradual and intertwined with the development of modern science and philosophy.

**Scientific Revolution (16th-17th Century):** The emergence of modern science, with figures like Galileo Galilei and Isaac Newton, challenged the static and hierarchical view of the universe. Scientific discoveries, such as the heliocentric model of the solar system and the laws of motion, provided alternative explanations for natural phenomena.

**Enlightenment Philosophy (18th Century):** Enlightenment thinkers questioned traditional authorities and embraced empiricism and reason. The Great Chain of Being, with its rigid hierarchy, began to seem increasingly untenable in the face of new ideas about human rights, equality, and individualism.

**Charles Darwin's Theory of Evolution (19th Century):** Charles Darwin's theory of evolution by natural selection was a pivotal moment in the decline of the Great Chain of Being. It challenged the idea of a fixed and unchanging hierarchy by proposing that species could change and adapt over time through a natural process.

### When the Great Chain of Being Was Disproved

The decline of the Great Chain of Being spanned several centuries, but it can be marked by significant milestones:

**17th Century:** The Scientific Revolution, characterized by the work of Galileo and Newton, began to challenge the traditional worldview.

**18th Century:** Enlightenment thinkers like Voltaire, Rousseau, and Kant critiqued the idea of a rigid cosmic hierarchy.

**19th Century:** Charles Darwin's theory of evolution, outlined in "On the Origin of Species" (1859), provided a comprehensive and evidence-based alternative to the Great

Chain of Being.

By the 19th century, the Great Chain of Being had largely lost its intellectual and philosophical standing, replaced by more dynamic and evidence-based views of the universe, life, and human society.

**Legacy and Insights**

The Great Chain of Being, while ultimately disproved, left a lasting legacy in the history of ideas. It offers insight into how pre-modern cultures sought to understand the natural and metaphysical world, emphasizing the importance of order, hierarchy, and the divine in shaping their worldview. While it may no longer hold sway in contemporary science and philosophy, the concept of the Great Chain of Being remains a fascinating chapter in the ongoing evolution of human thought and understanding.

# Lysenkoism

Lysenkoism stands as a stark example of the intertwining of science, politics, and ideology during the 20th century, particularly in the Soviet Union. This theory, named after its proponent Trofim Lysenko, veered far from established biological principles and led to significant consequences for Soviet science and agriculture. In this exploration, we will delve into the essence of Lysenkoism, its historical origins, key figures involved, its influence, the eventual discrediting of the theory, and the repercussions it had on Soviet biology.

**The Essence of Lysenkoism**

Lysenkoism was a collection of pseudoscientific ideas and practices in the field of biology that contradicted the principles of Mendelian genetics and Darwinian evolution. The central tenets of Lysenkoism included:

**Inheritance of Acquired Characteristics:** Lysenko rejected the idea of genetic inheritance and proposed that acquired traits could be passed on to offspring. This notion, known as Lamarckism, had been largely discredited in the scientific community by the 20th century.

**Vernalization:** Lysenko advocated for the practice of vernalization, where seeds or plants were subjected to cold temperatures to induce desirable traits. He claimed that this could increase crop yields and adapt crops to different climates.

**Denial of Natural Selection:** Lysenko rejected the concept of natural selection and believed that organisms could adapt rapidly to their environment through directed breeding and environmental conditioning.

### Origins of Lysenkoism

Lysenkoism emerged during a turbulent period in Soviet history, particularly during the 1930s and 1940s.

**Trofim Lysenko:** The theory is named after Trofim Lysenko, a self-taught biologist and agronomist. Lysenko rose to prominence within the Soviet scientific establishment due to his close alignment with the ideological views of Joseph Stalin, the leader of the Soviet Union.

### Proponents and Skeptics

During its heyday, Lysenkoism found support among high-ranking Soviet officials and was endorsed by the Communist Party. Key figures involved in promoting Lysenkoism included Joseph Stalin and Andrei Zhdanov.

**Critics and Skeptics:** Despite its official backing, Lysenkoism faced significant opposition within the scientific community. Prominent geneticists and biologists, such as Nikolai Vavilov and Georgii Karpechenko, criticized Lysenko's ideas and sought to uphold the principles of classical genetics.

### The Story of Discrediting Lysenkoism

The eventual discrediting of Lysenkoism was a complex and multifaceted process that unfolded over several decades.

**Suppression of Critics:** Lysenko and his supporters, backed by the Soviet government, actively suppressed and persecuted scientists who opposed Lysenkoism. This led to the silencing and sometimes imprisonment of geneticists and biologists who advocated for mainstream biological principles.

**Pseudoscientific Practices:** Despite the lack of empirical evidence, Lysenko's ideas were implemented in Soviet agriculture and biology. This resulted in disastrous consequences, including crop failures and food shortages.

**International Isolation:** The global scientific community largely rejected Lysenkoism, isolating Soviet biology from international research and collaboration. This isolation further hindered the development of Soviet science.

**Shift in Leadership:** After Joseph Stalin's death in 1953, the Soviet leadership gradually distanced itself from Lysenko. New leaders recognized the failures of Lysenkoist policies and began to reintegrate Soviet biology with international science.

### When Lysenkoism Was Disproved

The formal discrediting of Lysenkoism began in the late 1950s and continued throughout the 1960s.

**1964:** The Soviet Academy of Sciences officially rejected Lysenkoism and reinstated mainstream biological research.

**Late 20th Century:** Lysenkoism gradually faded from Soviet biology, and mainstream genetics and evolutionary biology were reintroduced into the curriculum and research.

### Legacy and Insights

Lysenkoism left a lasting legacy in the history of science and serves as a cautionary tale about the dangers of mixing ideology with scientific inquiry. It demonstrated the devastating consequences of subordinating scientific truth to political dogma and the suppression of dissenting voices. Lysenkoism's impact on Soviet agriculture also serves as a stark reminder of the importance of evidence-based practices in agriculture and biology. In the end, the rise and fall of Lysenkoism underscore the fragility of scientific integrity and the vital role of open and rigorous inquiry in advancing our understanding of the natural world.

# The Doctrine of Signatures

The Doctrine of Signatures, a fascinating yet obsolete concept, was a belief system rooted in the idea that nature carries hidden messages through the shapes, colors, and appearances of plants and other natural elements. While this theory may seem whimsical and mystical by modern standards, it held significant influence in the realms of medicine, botany, and philosophy during certain historical periods.

**Understanding the Doctrine of Signatures**

The Doctrine of Signatures, also known as the Law of Similarities, was a belief that nature provided clues to the medicinal properties of plants and substances through their physical characteristics. The core principles of this theory included:

**Similarity:** The doctrine posited that if a natural element resembled a particular part of the human body or had characteristics reminiscent of a certain ailment, it was a sign that this element held curative properties for that ailment.

**Divine Design:** Proponents of the doctrine believed that these signatures were not mere coincidences but rather a divine design meant to guide humans in discovering the healing properties of nature.

**Origins and Early Development**

The origins of the Doctrine of Signatures can be traced back to ancient civilizations, but it gained prominence during the Renaissance and the centuries that followed. Key points about its origins and development include:

**Ancient Roots:** The idea that nature's forms and colors held significance in medicine can be found in the writings of ancient civilizations, including the Greeks and Romans. These cultures observed and documented the use of plants in healing based on their appearances.

**Paracelsus:** The Renaissance-era Swiss physician and alchemist Paracelsus played a pivotal role in popularizing the Doctrine of Signatures. He believed that the external characteristics of plants indicated their internal properties and medicinal uses. Paracelsus' writings and teachings spread these ideas throughout Europe.

**Believers and Supporters**

The Doctrine of Signatures found supporters among naturalists, herbalists, and philosophers, including:

**Paracelsus:** As previously mentioned, Paracelsus was a prominent advocate of the doctrine. His writings, such as "Liber Paragranum," outlined his belief in the symbolic relationship between nature and healing.

**Herbalists:** Herbalists and folk healers often relied on the Doctrine of Signatures to identify plants with potential medicinal value. They believed that plants resembling body parts could treat ailments related to those parts.

**The Story of How the Theory Eventually Got Disproved**

The Doctrine of Signatures, though influential during its time, eventually fell out of favor and was disproved for several reasons:

**Empirical Observation:** As scientific methods and empirical observation became more prevalent, the reliance on symbolism and appearance-based medicine began to wane. Naturalists and botanists started to question the validity of the doctrine.

**Advancements in Medicine:** The development of modern medicine, with its emphasis on rigorous testing, clinical trials, and evidence-based practices, rendered the Doctrine of Signatures obsolete. Medical practitioners shifted away from relying on symbolism and instead sought concrete evidence of a substance's therapeutic efficacy.

**Understanding of Botany:** With advancements in the field of botany, scientists gained a more comprehensive understanding of plant biology and the chemical compounds responsible for medicinal properties. This knowledge allowed for a more accurate and rational approach to herbal medicine.

**When the Doctrine of Signatures Got Disproved**

The decline of the Doctrine of Signatures took place gradually over several centuries:

**Renaissance to Enlightenment:** The doctrine gained prominence during the Renaissance but began to decline during the Enlightenment period. Enlightenment thinkers and scientists advocated for evidence-based practices and questioned the superstitions associated with the doctrine.

**19th Century:** By the 19th century, the Doctrine of Signatures had largely faded from mainstream medicine and botany. It was no longer considered a valid or reliable approach to identifying medicinal plants or substances.

**Modern Era:** In the modern era, the doctrine is regarded as a historical curiosity rather than a scientific or medical framework. Modern herbal medicine relies on empirical evidence, chemical analysis, and clinical studies to determine the efficacy of medicinal plants and substances.

In conclusion, the Doctrine of Signatures was a belief system that attributed healing properties to the external appearances of natural elements. While it held sway during certain historical periods, it was eventually discredited as empirical observation, scientific advancements, and evidence-based medicine took precedence. Today, it serves as a reminder of humanity's evolving understanding of nature and medicine, highlighting the importance of critical thinking and empirical evidence in the field of healthcare.

# Ovism

Ovism, a historical theory of reproduction, proposed that all living organisms, including humans, originated from pre-existing eggs. This theory, which may appear peculiar from a modern scientific standpoint, had a significant influence on early biology and embryology. In this exploration of ovism, we will delve into its origins, key proponents, eventual disproof, and its place in the history of scientific thought.

**The Essence of Ovism**

Ovism, also known as the theory of preexistence, posited that complex life forms, including humans, were not generated from fertilization but instead developed from preformed eggs. This theory had several key components:

**Preexisting Eggs:** Ovism proposed that organisms, from the smallest insects to humans, had their origins in tiny, preformed eggs that existed within the female's body.

**No Role for Sperm:** Unlike the theory of spermism, which suggested that sperm contained miniature organisms, ovism entirely discounted the role of sperm in generating life. Instead, sperm was seen as merely a trigger for egg development.

**Homunculus:** Ovism occasionally included the concept of the homunculus, a miniature, fully-formed human present in the egg. This notion aligned with the idea that all the essential components of an organism were present from the beginning.

**The Emergence of Ovism**

The origins of ovism can be traced back to ancient Greece, where philosophers such as Empedocles and Anaximander speculated on the existence of preformed embryos. However, the theory gained prominence during the late 17th and early 18th centuries:

**Ancient Influences:** The idea of preexistence, although not fully developed as ovism, can be found in the works of ancient philosophers. Empedocles, for instance, believed in the existence of preformed embryos.

**17th Century:** Ovism began to take shape during the 17th century. Notably, Nicolaas Hartsoeker, a Dutch scientist, played a significant role in promoting the theory. In 1694, he claimed to have observed tiny, human-like homunculi in sperm using a microscope. This discovery further fueled the ovist perspective.

**18th Century:** Ovism continued to gain traction during the 18th century, with prominent scientists such as Caspar Friedrich Wolff advocating for the theory. Wolff proposed that each part of the body had its own specific egg.

**Proponents and Supporters**

Ovism found proponents among notable scientists and scholars of the time, including:

**Nicolaas Hartsoeker:** As mentioned earlier, Hartsoeker's observations with microscopes contributed significantly to the popularization of ovism.

**Caspar Friedrich Wolff:** Wolff, a prominent embryologist and naturalist, was a key advocate for ovism in the 18th century. He believed that eggs contained all the information needed for an organism's development.

### The Disproof of Ovism

Ovism eventually fell out of favor and was discredited for several reasons:

**Discovery of Fertilization:** The discovery of fertilization as a process, primarily attributed to Lazzaro Spallanzani in the late 18th century, provided a clear mechanism for the generation of new organisms. This directly contradicted the ovist perspective, which disregarded the role of sperm.

**Advancements in Microscopy:** While early microscopes hinted at the existence of tiny homunculi, more advanced microscopy techniques in the 19th century failed to provide concrete evidence for their existence.

**Developmental Biology:** The emergence of developmental biology as a field in the 19th century led to a deeper understanding of embryonic development. This scientific discipline highlighted the role of both male and female contributions in the formation of offspring.

### When Ovism Got Disproved

The downfall of ovism occurred gradually:

**Late 18th Century:** The disproof of ovism can be primarily attributed to Lazzaro Spallanzani's experiments on artificial insemination and fertilization in the late 18th century. His work demonstrated that sperm played a vital role in the generation of new life, contradicting the core tenets of ovism.

**19th Century:** Ovism gradually faded from scientific discourse during the 19th century as the field of embryology embraced fertilization as the primary mechanism for reproduction. The disproof of ovism marked a significant transition in our understanding of the origins of life.

In summary, ovism, the theory of preexistence, proposed

that all organisms, including humans, originated from pre-formed eggs and excluded the role of sperm in reproduction. While it had its proponents and influences in the history of biology, the theory was eventually discredited due to advancements in microscopy, the discovery of fertilization, and the development of developmental biology. Ovism serves as a reminder of the evolving nature of scientific knowledge and the importance of empirical evidence in shaping our understanding of the natural world.

# The Essence of Lamarckian Evolution

Lamarckian evolution, also known as Lamarckism, is a theory of biological evolution that posits that an organism can acquire new traits during its lifetime in response to environmental changes or needs. These acquired traits were believed to be inheritable, meaning that an organism could pass them on to its offspring. The central tenets of Lamarckian evolution can be summarized as follows:

**Use and Disuse:** Lamarck proposed that an organism could develop new traits through the use or disuse of specific organs or body parts. For example, if a giraffe stretched its neck to reach leaves high in the trees, its offspring would inherit longer necks.

**Inheritance of Acquired Characteristics:** Lamarck argued that the traits acquired by an organism during its lifetime would be passed on to its offspring. This idea contrasted with the prevailing belief in the early 19th century, which was a time when Lamarck's ideas gained prominence.

**Origins of Lamarckian Evolution**

**Jean-Baptiste Lamarck:** The theory is named after Jean-Baptiste Lamarck, a French biologist and naturalist. Lamarck first proposed his ideas in the early 19th century, particularly in his work "Philosophie Zoologique" published in 1809.

**Prominent Figures and Believers**

Lamarck's ideas found both supporters and critics during his time.

**Supporters:** Lamarckian evolution gained traction in the early 19th century and found support among some notable scientists and naturalists. Étienne Geoffroy Saint-Hilaire and Geoffroy's student, Étienne Serres, were among those who embraced Lamarck's ideas.

**Critics:** While Lamarck's theory gained attention, it also faced criticism from prominent scientists of the time, including Georges Cuvier and Charles Lyell. Cuvier's extensive work in paleontology and comparative anatomy did not align with Lamarckism, and he argued against the concept of the inheritance of acquired characteristics.

**The Story of How Lamarckian Evolution Got Disproved**

The eventual discrediting of Lamarckian evolution was closely tied to the emergence of Charles Darwin's theory of natural selection.

**Publication of "On the Origin of Species" (1859):** Charles Darwin's seminal work presented a radically different theory of evolution based on natural selection. Darwin's theory emphasized the role of heritable variation and differential reproductive success in shaping populations over time. This stood in stark contrast to Lamarck's ideas.

**Mendel's Laws of Inheritance:** Gregor Mendel's experiments with pea plants in the mid-19th century provided strong evidence for the particulate nature of inheritance. Mendel's laws, which were rediscovered in the early 20th century, demonstrated the role of discrete units (genes) in heredity, rather than the inheritance of acquired traits as proposed by Lamarck.

**Modern Synthesis:** In the 20th century, the modern synthesis of evolutionary biology integrated Mendelian genetics with natural selection, effectively replacing Lamarckian evolution as the dominant paradigm in biology. This synthesis provided a comprehensive framework for understanding how populations evolve over time.

### When Lamarckian Evolution Was Disproved

Lamarckian evolution gradually lost credibility throughout the 19th and 20th centuries.

**Mid-19th Century:** Charles Darwin's theory of natural selection and Gregor Mendel's work on genetics began to eclipse Lamarckian evolution.

**Early 20th Century:** The rediscovery and acceptance of Mendel's laws of inheritance, combined with the development of population genetics, further discredited Lamarckism.

**Modern Synthesis (1930s-1940s):** The modern synthesis, which merged Mendelian genetics with the theory of natural selection, marked the formal rejection of Lamarckian evolution by the scientific community.

### Legacy and Insights

Lamarckian evolution, despite its eventual dismissal, played a significant role in the history of evolutionary thought. It highlighted the early attempts to explain the mechanisms of evolution and adaptation before the discovery of the genetic basis of inheritance. Lamarck's ideas also served as a foil against which Charles Darwin's theory of natural selection emerged as a more robust and widely accepted explanation for the diversity of life on Earth.

Today, Lamarckism is regarded as a historical curiosity rather than a valid scientific theory. It serves as a reminder of the importance of empirical evidence and rigorous scientific scrutiny in shaping our understanding of the natural world. While Lamarck's proposal of the inheritance of acquired traits may have seemed plausible in his time, the subsequent advancements in genetics and evolutionary biology have revealed a more complex and accurate picture of how species change and adapt over time.

# The Panspermia Hypothesis

The Panspermia Hypothesis is a fascinating scientific theory proposing that life, in the form of microorganisms or organic molecules, can exist and travel through space, eventually seeding other celestial bodies like planets. This concept has captivated scientists and thinkers for centuries, offering a unique perspective on the origins and distribution of life in the universe.

### Origins of the Panspermia Hypothesis

The concept of panspermia has ancient roots, with early speculations suggesting that life could have extraterrestrial origins. However, the modern Panspermia Hypothesis, as we understand it today, emerged in the 5th century BC with the Greek philosopher Anaxagoras. He proposed that life could have originated from other worlds and was brought to Earth by celestial impacts.

### Anaxagoras: The Early Proponent

Anaxagoras is credited as one of the earliest proponents of the Panspermia Hypothesis. He believed that life was ubiquitous in the cosmos and that it could be transported by meteorites or comets, which served as carriers of life from one celestial body to another.

### Key Supporters of Panspermia

Over the centuries, several notable scientists and thinkers have supported the Panspermia Hypothesis:

**Hermann von Helmholtz:** In the 19th century, the German physicist and physician Hermann von Helmholtz proposed that spores of life could exist in interstellar space and be transported across vast distances by radiation pressure.

**Lord Kelvin:** The renowned British physicist Lord Kelvin suggested that life might have originated on other planets and been transported to Earth by meteorites.

**Sir Fred Hoyle and Chandra Wickramasinghe:** In the 20th century, the astrophysicist Sir Fred Hoyle and the mathematician Chandra Wickramasinghe championed the modern Panspermia Hypothesis. They argued that complex organic molecules and even microorganisms could travel through space on dust particles and eventually settle on planets.

**Scientific Evidence and Disproof**

While the Panspermia Hypothesis remains a compelling idea, it has not been conclusively proven. Scientific evidence and technological advancements have shed light on its feasibility:

**Meteorites:** The discovery of organic molecules, amino acids, and even water in certain meteorites has provided some support for the idea that life's building blocks could exist in space.

**Extremophiles:** The discovery of extremophiles, organisms that thrive in extreme environments on Earth, has expanded our understanding of where life can exist and has raised the possibility that similar life forms could exist elsewhere.

**Experiments in Space:** Experiments conducted in space

have demonstrated that certain microorganisms can survive and even thrive in the harsh conditions of space, suggesting that panspermia is theoretically possible.

**Limits to Panspermia:** Despite these intriguing findings, the Panspermia Hypothesis faces significant challenges. The extreme conditions of space, such as high levels of radiation and vacuum, make the survival of unprotected microorganisms unlikely during interstellar journeys.

## Conclusion and Ongoing Exploration

The Panspermia Hypothesis, which suggests that life can travel through space and seed other celestial bodies, has ancient origins but gained prominence with the contributions of thinkers like Anaxagoras and scientists like Lord Kelvin, Hermann von Helmholtz, and Fred Hoyle. While it remains an intriguing concept, advances in astrobiology, the study of extremophiles, and experiments in space have raised questions about the feasibility of panspermia.

As our understanding of the cosmos continues to evolve and our exploration of other planets and moons expands, panspermia remains a topic of scientific interest and investigation. While it has not been conclusively proven, it offers a unique perspective on the potential for life beyond Earth and the interconnectedness of the universe. The Panspermia Hypothesis challenges us to consider the possibility that the cosmic seeds of life may have traveled through the vast expanse of space, shaping the destiny of worlds far beyond our own.

# Pangenesis

Pangenesis, a historical theory of heredity, proposed that the body's cells shed tiny particles called "gemmules" that carried hereditary information. These gemmules would collect in the reproductive organs and then be passed on to the next generation, influencing the traits of offspring. Pangenesis, although ultimately proven incorrect, played a significant role in the development of genetics and our understanding of inheritance.

**The Essence of Pangenesis**

The theory of pangenesis had several key components:

**Gemmules:** Pangenesis proposed the existence of gemmules, minute particles produced by cells throughout an organism's body. These gemmules carried information about the organism's traits.

**Reproductive Collection:** According to pangenesis, these gemmules would travel to the organism's reproductive organs, where they would accumulate.

**Inheritance:** During reproduction, the gemmules from both parents would combine to form the traits of the offspring. This process was believed to account for the inheritance of acquired characteristics.

**Origins of Pangenesis**

The origins of pangenesis can be traced back to ancient Greece, where philosophers such as Hippocrates and Aristotle pondered the nature of heredity. However, the fully developed theory of pangenesis emerged during the 19th century:

**Greek Philosophers:** Ancient Greek philosophers like Hippocrates and Aristotle made early observations about heredity, but their ideas were rudimentary compared to pangenesis.

**19th Century:** The modern theory of pangenesis began to take shape in the 19th century. Charles Darwin, the renowned naturalist, played a pivotal role in popularizing pangenesis. In his book "The Variation of Animals and Plants under Domestication" (1868), Darwin proposed the theory to explain how acquired characteristics could be inherited.

### Proponents and Supporters

Pangenesis found proponents among prominent scientists and scholars of the time, including:

**Charles Darwin:** As the most famous proponent of pangenesis, Charles Darwin included this theory in his work "The Variation of Animals and Plants under Domestication." Darwin believed that gemmules could account for the inheritance of acquired traits, a concept he considered essential to his theory of evolution.

### The Disproof of Pangenesis

Pangenesis eventually faced challenges and was disproved for several reasons:

**Lack of Evidence:** One of the main criticisms of pangenesis was the lack of empirical evidence for the existence of gemmules. Despite extensive efforts to observe or isolate these particles, they were never discovered.

**Mendel's Laws:** Gregor Mendel's groundbreaking experiments with pea plants, published in the mid-19th century but largely overlooked until the early 20th century, provided a more accurate and scientifically sound explanation of inheritance. Mendel's laws of inheritance introduced the concept of discrete units of heredity, which we now know as genes.

**When Pangenesis Got Disproved**

Pangenesis was gradually discredited during the late 19th and early 20th centuries:

**Mid-19th Century:** Gregor Mendel's experiments on pea plants, which demonstrated the existence of discrete units of heredity (genes), began to cast doubt on the idea of gemmules and pangenesis.

**Early 20th Century:** As Mendel's laws gained recognition and support from empirical evidence, pangenesis lost favor among scientists. The discovery of chromosomes and the identification of DNA as the genetic material provided further evidence against pangenesis.

In summary, pangenesis was a historical theory of heredity that proposed the existence of gemmules, tiny particles carrying hereditary information, which were shed by cells throughout an organism's body. These gemmules were believed to collect in the reproductive organs and influence the traits of offspring. Although initially championed by Charles Darwin, the theory lacked empirical evidence and was eventually disproved with the emergence of Mendel's laws of inheritance and the discovery of genes and DNA. Pangenesis serves as a testament to the evolving nature of scientific understanding, as it paved the way for the development of modern genetics.

# The Spontaneous Generation of Mice

The theory of the spontaneous generation of mice is a captivating example of an erroneous scientific idea that persisted for centuries. This theory posited that mice could spontaneously generate from non-living matter, such as decaying organic material, and was widely accepted in various forms throughout history. Here, we delve into the origins, proponents, disproof, and eventual rejection of this curious belief.

## Origins of the Theory

The concept of spontaneous generation dates back to ancient times and was rooted in observations of seemingly "spontaneously" generated life forms. Aristotle, the ancient Greek philosopher, wrote about this idea in his work "History of Animals" (c. 350 BC). He proposed that some animals, like insects and certain small creatures, could arise from non-living matter under specific conditions. Aristotle's influence was so profound that his ideas shaped scientific thought for centuries.

## Aristotle's Influence

Aristotle's ideas on spontaneous generation influenced many prominent figures in the centuries that followed. In the Roman era, scholars like Pliny the Elder believed in the spontaneous generation of various animals, including mice. Pliny's influential work, "Naturalis Historia" (c. 77 AD), included descriptions of spontaneous generation.

## The Middle Ages and Early Modern Period

The theory of spontaneous generation persisted through the Middle Ages and into the early modern period. During these times, it was widely accepted that mice, as well as other creatures like flies and maggots, could arise from decaying organic matter. This belief was often supported by anecdotal evidence, as people observed mice near food sources and assumed they had spontaneously generated from the food.

## Francesco Redi's Experiment

The eventual disproof of the spontaneous generation of mice and other animals can be attributed to the Italian physician and biologist Francesco Redi. In the 17th century, Redi conducted a groundbreaking experiment that challenged the prevailing belief.

## The Redi Experiment (1668)

Francesco Redi designed an experiment to test whether maggots, thought to spontaneously generate on decaying meat, could actually arise from pre-existing eggs. He placed meat in several jars, covering some with fine gauze and leaving others uncovered. In the uncovered jars, flies could access the meat and lay their eggs, while the gauze-covered jars prevented flies from reaching the meat.

Over time, Redi observed that maggots only appeared in the uncovered jars, where flies had access. The gauze-covered jars remained free of maggots. This experiment provided strong evidence against the spontaneous generation of maggots and, by extension, mice and other animals.

## The Role of Francesco Redi

Redi's experiment marked a significant turning point in the scientific understanding of life's origins. His work challenged the long-held belief in spontaneous generation, demonstrating that living organisms could only come from pre-existing life forms. While Redi's experiment primarily targeted maggots, it had broader implications for the theory of spontaneous generation as a whole.

## Louis Pasteur's Contribution

Although Redi's experiment dealt a substantial blow to the idea of spontaneous generation, the theory continued to linger in certain forms. It wasn't until the 19th century that the final nail in the coffin was driven by the renowned French chemist and microbiologist Louis Pasteur.

## The Swan Neck Flask Experiment (1862)

Pasteur's famous experiment involved swan-necked flasks filled with a nutrient broth. The flasks had long, curved necks that allowed air to enter but prevented particles from reaching the broth. Pasteur boiled the broth to sterilize it and then left the flasks exposed to the air.

Over time, Pasteur observed that no microorganisms appeared in the broth. Only when he tilted the flasks, allowing particles from the curved neck to enter the broth, did microbial growth occur. This experiment conclusively demonstrated that microorganisms could not spontaneously generate in a sterile environment but required the introduction of outside particles.

## The Decline of Spontaneous Generation

With Pasteur's experiment and the growing understanding of microbiology, the theory of spontaneous generation lost all scientific credibility. By the mid-19th century, it was firmly established that life could only arise from pre-existing life forms. The idea that mice, or any other complex organisms, could spontaneously generate from non-living matter had been thoroughly discredited.

## Conclusion

The theory of the spontaneous generation of mice and other animals was a deeply ingrained belief that spanned centuries. It originated with Aristotle's ancient observations and persisted through the Middle Ages and early modern period. It wasn't until the experiments of scientists like Francesco Redi and Louis Pasteur in the 17th and 19th centuries that the theory was decisively disproven.

Redi's experiments with maggots and Pasteur's work with microorganisms provided the empirical evidence needed to reject the idea of spontaneous generation. These experiments marked critical milestones in the history of science, leading to the development of the biogenesis theory, which states that life arises only from pre-existing life. The decline of the spontaneous generation theory represented a significant step forward in our understanding of the origins of life and contributed to the foundation of modern biology.

# The Doctrine of Signature Medicines

The Doctrine of Signature Medicines, a theory that dates back to ancient times, was a belief system that suggested a connection between the appearance of plants and their medicinal properties. It proposed that the physical characteristics of plants, such as their shape, color, or texture, could provide clues about their healing properties. This theory gained popularity during the Middle Ages and persisted for centuries. In this exploration, we will delve into the origins of the Doctrine of Signature Medicines, its proponents, its eventual decline in credibility, and when it was disproved.

**What Was the Doctrine of Signature Medicines?**

The Doctrine of Signature Medicines, also known as the Law of Signatures, was a belief system that linked the physical attributes of plants to their potential medicinal uses. According to this theory, the appearance of a plant held clues about its healing properties. This connection was believed to be established by a divine creator, suggesting that nature provided signs for humans to discover the therapeutic benefits of various plants.

For example, plants with heart-shaped leaves were thought to be useful for heart-related ailments, while plants with yellow sap or blossoms were associated with the treatment of jaundice or liver disorders. The idea was that nature had "signed" these plants with visual cues, making it easier for humans to identify their medicinal uses.

## Origins and Early Proponents

The Doctrine of Signature Medicines has ancient roots and can be traced back to various cultures and civilizations. However, it gained prominence in Europe during the Middle Ages. Key figures associated with the theory include:

**Paracelsus (1493-1541):** A Swiss physician, alchemist, and astrologer, Paracelsus is often cited as one of the early proponents of the Doctrine of Signature Medicines. He believed that the properties of plants could be discerned from their external characteristics.

**Jakob Böhme (1575-1624):** A German mystic and philosopher, Böhme expanded on the concept, suggesting that there were hidden spiritual meanings within the natural world, including plants.

## Scientific Reception and Challenges

The Doctrine of Signature Medicines gained popularity during a time when scientific understanding was limited. It appealed to the human desire to find order and meaning in the natural world. However, as scientific knowledge advanced, several challenges and criticisms emerged:

**Lack of Empirical Evidence:** The theory was based on subjective observations and lacked empirical evidence to support its claims. It relied on anecdotal associations rather than rigorous scientific experimentation.

**Inconsistencies:** The Doctrine of Signature Medicines often led to inconsistencies and contradictions. Plants with similar appearances were attributed to different medicinal properties, and the theory could not account for the complexity of plant chemistry.

**Advancements in Medicine:** As modern medicine developed, it became evident that the therapeutic value of plants was not solely determined by their external characteristics. Chemical analysis and pharmacological studies revealed the diverse and complex compounds within plants that contributed to their medicinal properties.

## Decline and Disproof

The decline of the Doctrine of Signature Medicines can be attributed to the advancements in scientific understanding and the emergence of evidence-based medicine. The theory's reliance on symbolism and superficial observations became increasingly untenable in the face of rigorous scientific inquiry.

While it is challenging to pinpoint an exact moment when the theory was definitively disproved, its decline in credibility occurred over several centuries. The emergence of the scientific method and the development of pharmacology provided more accurate and reliable ways to study the medicinal properties of plants.

By the 19th and 20th centuries, the Doctrine of Signature Medicines had largely been relegated to the realm of folklore and historical curiosity. It no longer held sway in mainstream medical practice or scientific discourse. Instead, the field of pharmacognosy, which focuses on the study of medicinal compounds in plants, became the foundation for modern herbal medicine.

In conclusion, the Doctrine of Signature Medicines was a belief system that linked the physical attributes of plants to their potential medicinal uses. It gained popularity during the Middle Ages but gradually lost credibility as scientific knowledge advanced. The lack of empirical evidence, inconsistencies, and the emergence of evidence-based medicine contributed to its decline. While it is challenging to pinpoint a specific moment of disproof, the theory's decline occurred over several centuries, and it is no longer a relevant concept in contemporary medicine.

## Vitalism: The Once-Influential Theory of Vital Life Force

Vitalism was a prominent scientific and philosophical theory that persisted for centuries and proposed the existence of a vital life force or energy that was responsible for the essential characteristics of living organisms. This theory, which originated in ancient times and continued through the early modern period, attracted the attention of many prominent scientists and thinkers. However, it gradually lost credibility as modern science advanced, particularly with the advent of biochemistry and molecular biology. In this exploration, we will delve into the essence of vitalism, its historical origins, its early proponents, the influential individuals who believed in its validity, the process by which it was ultimately disproved, and the timeline of its discrediting.

### The Essence of Vitalism

Vitalism posited that living organisms possessed a unique and indefinable vital force or energy that distinguished them from inanimate matter. This vital force was believed to be responsible for the processes of life, growth, and organization that could not be explained solely by physical and chemical principles.

### Origins of Vitalism

The origins of vitalism can be traced back to ancient beliefs in the existence of a vital life force.  Many early cultures, including the Greeks and Egyptians, held beliefs in the soul or spirit as the animating principle of life.

**Ancient Notions of Vital Force:**  In ancient Greece, philosophers such as Aristotle explored the concept of a vital force or "anima" that animated living organisms. Similar ideas were present in other ancient cultures, where life was often associated with a unique, immaterial force.

### Proponents of Vitalism

Throughout history, vitalism attracted the attention of numerous scientists, philosophers, and thinkers who believed in the existence of a vital life force.

**Rene Descartes' Animal Spirits:** René Descartes, the renowned French philosopher and mathematician, proposed a form of vitalism in the 17th century.  He introduced the concept of "animal spirits," which were believed to be vital forces that flowed through the nerves and animated the body.

**Early Biologists' Vitalism:** In the 18th century, vitalism influenced early biologists who believed that living organisms were distinct from non-living matter due to the presence of a vital force.

### The Decline of Vitalism: Emergence of Mechanistic Science

The credibility of vitalism began to decline as the scientific method gained prominence, and researchers sought natural explanations for the phenomena of life. Several key developments contributed to the disproof of this theory.

**Mechanistic Biology:** The emergence of mechanistic

biology in the 17th and 18th centuries emphasized the idea that living organisms could be understood through physical and chemical principles, rather than relying on an indefinable vital force.

**Chemistry and Biochemistry:** Advances in chemistry and biochemistry revealed the chemical composition and processes underlying life. Vitalism's inability to explain these chemical processes further eroded its credibility.

**Louis Pasteur and Germ Theory:** Louis Pasteur's groundbreaking work on fermentation and the germ theory of disease provided a scientific basis for understanding many biological processes. His experiments demonstrated that microorganisms, rather than a vital force, were responsible for fermentation and disease.

**When Vitalism Was Disproved**

The discrediting of vitalism occurred over a span of centuries, but several key developments marked significant milestones in its decline.

**19th-Century Advances in Physiology:** In the 19th century, advances in physiology provided detailed explanations for many biological functions, further undermining vitalism.

**Synthesis of Urea (1828):** The synthesis of urea by Friedrich Wöhler in 1828 is often cited as a pivotal moment in the decline of vitalism. Urea, previously thought to be a product of vital processes, was successfully synthesized from inorganic chemicals, challenging the idea that only living organisms could produce it.

**Arrival of Biochemistry (20th Century):** The emergence of biochemistry in the 20th century provided a comprehensive understanding of the biochemical processes underlying life. Biochemical reactions and pathways could be explained without invoking a vital force.

**Legacy and Insights**

The decline of vitalism underscores the importance of empirical evidence, the scientific method, and naturalistic explanations in understanding the phenomena of life. While vitalism was once influential, it eventually gave way to a mechanistic and materialistic view of biology. Modern biology, including genetics, molecular biology, and biochemistry, has flourished under this paradigm, providing profound insights into the workings of living organisms. Vitalism remains a historical curiosity, serving as a reminder of the evolving nature of scientific understanding and the importance of critical inquiry and evidence-based reasoning in the pursuit of knowledge.

*Lamarckism: The Evolutionary Theory That Time Left Behind*

Lamarckism, an early theory of evolution, once held a prominent place in the scientific discourse of its time. It proposed that organisms could acquire new traits during their lifetime and pass these acquired characteristics to their offspring. Although this theory was groundbreaking in its era, it eventually gave way to the more comprehensive and evidence-supported theory of natural selection proposed by Charles Darwin. In this exploration, we will dive into what Lamarckism entailed, when it emerged, who first conceived it, the notable figures who embraced it, the story of its decline, and when it was eventually disproven.

## The Essence of Lamarckism

Lamarckism, named after the French naturalist Jean-Baptiste Lamarck, is a theory of evolution that posits two key principles:

**The Inheritance of Acquired Characteristics:** Lamarck proposed that organisms could change during their lifetimes in response to their environments, and these changes could be passed on to their offspring. For example, if a giraffe stretched its neck to reach high leaves, it would develop a longer neck, and this acquired trait would be inherited by its descendants.

**Use and Disuse:** Lamarck suggested that organisms could gain or lose traits based on how much they used or didn't use them. Organs or characteristics that were used extensively would become more developed, while those that were not used would deteriorate over generations.

## The Birth of Lamarckism: Its Origin

Lamarckism emerged in the late 18th and early 19th centuries, a time when the study of biology and the concept of evolution were in their infancy. Jean-Baptiste Lamarck, a French naturalist and biologist, is credited with formalizing the theory in his work "Philosophie Zoologique," published in 1809.

## Lamarck and His Advocates: The Believers

During its heyday, Lamarckism attracted the support of notable scientists and thinkers. Lamarck's ideas resonated with those who sought an explanation for the diversity of life on Earth. Some of his contemporaries and followers who embraced Lamarckism included Étienne Geoffroy Saint-Hilaire and Jean-Baptiste de Monet, Chevalier de Lamarck's theory seemed to provide a mechanism for how species could adapt and change over time.

**The Rise and Fall of Lamarckism: A Story of Scientific Revolution**

The eventual decline of Lamarckism can be attributed to several factors, including a lack of empirical evidence and the emergence of competing evolutionary theories, most notably Charles Darwin's theory of natural selection. Darwin's theory, outlined in "On the Origin of Species" in 1859, provided a more comprehensive and evidence-based explanation for the diversity of life.

**Darwin's Natural Selection:** Darwin's theory of natural selection proposed that individuals within a species vary in traits, and those with traits better suited to their environment are more likely to survive and reproduce. Over time, these advantageous traits become more common in the population through a process of "survival of the fittest."

**The Demise of Lamarckism: Disproving Acquired Characteristics**

Lamarck's central idea of the inheritance of acquired characteristics faced numerous challenges. One critical issue was the lack of empirical evidence to support the idea that traits acquired during an organism's lifetime could be passed on to its offspring. While some examples of short-term adaptation were observed in nature, such as changes in beak size in response to food availability, these did not provide conclusive proof of Lamarckian inheritance.

Furthermore, as genetics advanced and our understanding of heredity improved, it became clear that the mechanisms behind inheritance did not align with Lamarckism. Gregor Mendel's work on the principles of heredity, published in the mid-19th century but largely unrecognized at the time, laid the foundation for our modern understanding of genetics. Mendel's research demonstrated that traits were inherited through discrete units (later known as genes) and did not change during an organism's lifetime.

## The Dawn of Modern Evolutionary Biology: The End of Lamarckism

By the late 19th century, Lamarckism had lost its appeal among the scientific community. Charles Darwin's theory of natural selection, supported by emerging genetic knowledge, provided a more comprehensive and evidence-based explanation for the diversity of life on Earth. Natural selection offered a mechanism for the gradual evolution of species through the differential survival and reproduction of individuals with advantageous traits.

In essence, Lamarckism was gradually replaced by the modern synthesis of evolutionary biology, which integrated Darwinian natural selection with Mendelian genetics. This synthesis, which emerged in the early 20th century, provided a robust framework for understanding how species evolve and change over time.

In summary, Lamarckism, an early theory of evolution proposed by Jean-Baptiste Lamarck in the early 19th century, suggested that organisms could acquire new traits during their lifetimes and pass them on to their offspring. While it attracted some support during its time, Lamarckism eventually faced challenges due to a lack of empirical evidence and was superseded by Charles Darwin's theory of natural selection and the modern synthesis of evolutionary biology. This transition marked a pivotal moment in the history of science, leading to our contemporary understanding of evolution and heredity.

# Allopatric Speciation

Allopatric speciation is a fundamental concept in the field of evolutionary biology, explaining how new species can emerge from a common ancestor when populations become geographically isolated. This process involves the accumulation of genetic differences over time, driven by natural selection and genetic drift. Allopatric speciation has been a cornerstone of evolutionary theory, contributing to our understanding of biodiversity and the mechanisms underlying the origin of species. In this exploration, we will delve into the essence of allopatric speciation, its historical origins, notable proponents, the scientific consensus surrounding the theory, the evidence supporting it, and the contemporary understanding of this process.

**The Essence of Allopatric Speciation**

Allopatric speciation, often referred to as geographic speciation, occurs when populations of a single species become physically isolated from each other by geographical barriers. This isolation prevents gene flow between the separated populations, leading to divergence over time. The key elements of allopatric speciation include:

**Geographic Isolation:** Populations of a species are separated by geographical barriers such as mountains, rivers, or bodies of water.

**Genetic Divergence:** Isolated populations accumulate genetic differences through mutation, genetic drift, and natural selection.

**Reproductive Isolation:** Over time, these genetic differences can lead to reproductive barriers, preventing interbreeding between the separated populations.

**Emergence of New Species:** When reproductive isolation is complete, the two populations are considered separate species.

### Origins of the Allopatric Speciation Theory

The concept of allopatric speciation can be traced back to the mid-20th century, and it emerged as an integral part of evolutionary biology.

**Ernst Mayr:** One of the key figures in the development of the allopatric speciation theory was Ernst Mayr, a German-American biologist. Mayr's work, particularly in his 1942 book "Systematics and the Origin of Species," highlighted the role of geographical isolation in the formation of new species. He argued that isolation was a critical factor in the divergence of populations.

### Proponents and Scientific Consensus

Allopatric speciation has gained widespread acceptance among scientists and is now considered one of the primary mechanisms of species formation.

**Ernst Mayr:** Mayr's work was instrumental in establishing the concept of allopatric speciation. His contributions to the field of evolutionary biology have left a lasting legacy, and his ideas have been influential in shaping our understanding of the origin of species.

### Evidence Supporting Allopatric Speciation

Several lines of evidence support the concept of allopatric speciation:

**Geographical Patterns:** Observations of species distribution often reveal patterns consistent with geographic isolation leading to speciation. Islands, for example, often host unique species that have evolved in isolation.

**Genetic Divergence:** Genetic studies have shown that populations separated by geographical barriers tend to accumulate genetic differences over time. This genetic divergence can be seen in the form of DNA sequence differences, morphological variation, and reproductive incompatibilities.

**Experimental Studies:** Laboratory experiments and studies in the field have demonstrated that when populations are physically separated, they can undergo genetic divergence, leading to the development of reproductive barriers.

### Contemporary Understanding

The concept of allopatric speciation remains a fundamental principle of evolutionary biology. While it is widely accepted, researchers continue to explore the complexities of speciation, including the interplay between geography, genetics, and ecology.

**Challenges and Ongoing Research:** Scientists are studying cases of speciation in progress and investigating how ecological factors and natural selection interact with geographical isolation to shape the process of speciation.

**Integration with Other Mechanisms:** Allopatric speciation is often considered alongside other mechanisms of speciation, such as sympatric speciation, which occurs without geographical separation. Understanding how these mechanisms can interact is an ongoing area of research.

**When the Allopatric Speciation Theory Was Validated**

Allopatric speciation has been supported by a wealth of evidence and is a foundational concept in modern evolutionary biology. It has not been disproved but has continued to evolve and expand in light of new research and discoveries.

**Legacy and Insights**

Allopatric speciation stands as a testament to the power of geographical isolation in driving evolutionary change. It has provided a framework for understanding how the diversity of life on Earth has arisen over geological time. The theory has enriched our comprehension of the dynamic processes that lead to the emergence of new species and the incredible variety of life forms we observe today. In essence, allopatric speciation exemplifies the elegance of nature's mechanisms for generating biodiversity and continues to inspire curiosity and exploration in the field of evolutionary biology.

*Recapitulation Theory: The Controversial Idea of Ontogeny Recapitulating Phylogeny*

Recapitulation theory, also known as the theory of ontogeny recapitulating phylogeny, was a scientific hypothesis in the field of biology that proposed a connection between the development of an individual organism (ontogeny) and the evolutionary history of its species (phylogeny). According to this theory, the stages of an individual organism's development would mirror or recapitulate the evolutionary history of its species. While recapitulation theory had a long and influential history, it eventually faced scientific scrutiny and challenges. It is important to note that recapitulation theory is now largely discredited, but it remains a significant chapter in the history of biology. In this exploration, we will delve into the essence of recapitulation theory, its historical origins, its early proponents, the influential individuals who believed in its validity, the process by which it was ultimately disproved, and the timeline of its discrediting.

## The Essence of Recapitulation Theory

Recapitulation theory, also referred to as the biogenetic law, posited that during the development of an individual organism, its embryonic and fetal stages would pass through stages that reflected the evolutionary history of its species. This theory suggested that the embryonic development of an organism would resemble the adult forms of its evolutionary ancestors.

## Origins of Recapitulation Theory

The origins of recapitulation theory can be traced back to the 18th century, but it gained prominence and underwent significant development in the 19th century, coinciding with advances in the study of embryology and the theory of evolution.

**Étienne Geoffroy Saint-Hilaire:** The French naturalist Étienne Geoffroy Saint-Hilaire was one of the early proponents of recapitulation theory. In the early 19th century, he proposed that the development of an organism's embryo would mirror the stages of evolution experienced by its species.

**Karl Ernst von Baer:** The Estonian biologist Karl Ernst von Baer contributed to the understanding of embryology by emphasizing the distinctiveness of embryonic development. While his work laid the groundwork for modern embryology, he did not fully embrace recapitulation theory.

## Proponents of Recapitulation Theory

Recapitulation theory had influential proponents within the scientific community of the 19th and early 20th centuries, including Ernst Haeckel and others.

**Ernst Haeckel:** The German biologist Ernst Haeckel is perhaps the most well-known advocate of recapitulation theory. He coined the famous phrase "ontogeny recapitulates phylogeny" and produced illustrations known as "biogenetic trees" to illustrate the supposed relationship between embryonic development and evolutionary history.

**Louis Agassiz:** The Swiss-American biologist Louis Agassiz also supported recapitulation theory to some extent. He saw similarities between the embryonic stages of fish and the adult forms of other vertebrates.

## The Decline of Recapitulation Theory: Scientific Scrutiny and Challenges

Recapitulation theory began to face challenges and criticism as scientific understanding advanced, particularly in the fields of embryology and genetics.

**Developmental Variability:** As scientists gained a better understanding of the variability in embryonic development, it became clear that not all organisms followed the same developmental path. This variability contradicted the strict predictions of recapitulation theory.

**Genetics and Molecular Biology:** The advent of genetics and molecular biology in the 20th century provided more precise explanations for the development of organisms. The discovery of DNA and the understanding of how genes control development undermined the simplistic ideas of recapitulation theory.

**Empirical Observations:** Careful empirical observations of embryonic development in various species did not consistently support the idea that ontogeny recapitulated phylogeny.

### When Recapitulation Theory Was Disproved

Recapitulation theory faced increasing challenges and criticisms throughout the 20th century, ultimately leading to its discrediting.

**Late 19th to Early 20th Century:** Scientists like Karl Ernst von Baer and Wilhelm His Sr. had already cast doubt on recapitulation theory by emphasizing the unique aspects of embryonic development.

**Mid-20th Century:** With the advancement of genetics and the molecular understanding of development, recapitulation theory became less relevant and less accepted within the scientific community.

### Legacy and Insights

While recapitulation theory has been largely discredited, it played a significant role in the history of biology and the development of evolutionary thought. It highlights the importance of empirical evidence and the need for scientific ideas to evolve and adapt as new knowledge emerges. Despite its ultimate rejection, recapitulation theory contributed to the broader understanding of embryology and the relationships between development and evolution. It serves as a reminder that scientific theories, no matter how influential, are subject to scrutiny and revision in the pursuit of more accurate explanations of the natural world.

# The Theory of Maternal Impression

The theory of maternal impression is a historical belief that dates back centuries, asserting that a pregnant woman's thoughts, experiences, or emotions can directly influence the physical characteristics or well-being of her unborn child. In this exploration, we will delve into the origins of this theory, its proponents, the eventual disproof of its claims, and the impact it had on historical perceptions of pregnancy and childbirth.

### Origins of the Theory

The theory of maternal impression finds its roots in ancient beliefs about the profound connection between a pregnant woman and her developing fetus. It was believed that a pregnant woman's experiences, particularly her emotional and psychological state, could imprint directly onto her unborn child, affecting its physical appearance, health, and even temperament.

### Ancient Beliefs and Influences

Ancient cultures across the world held various beliefs related to maternal impressions. In ancient Egypt, for instance, it was thought that the emotions and experiences of pregnant women could shape the characteristics of their offspring. This belief was not limited to a specific culture but rather a widespread notion rooted in the limited scientific understanding of reproduction during ancient times.

## Hippocrates and the Ancient Greeks

The ancient Greeks, notably the famed physician Hippocrates (c. 460-370 BC), played a significant role in shaping the theory of maternal impression. Hippocrates believed in the importance of a woman's mental and emotional state during pregnancy and its potential impact on the child's development. His writings on this topic, including the influential "On the Nature of the Child," further disseminated the idea.

## Continuation Through the Middle Ages and Renaissance

The belief in maternal impression continued to be influential during the Middle Ages and the Renaissance. It was widely held that a pregnant woman's emotions, desires, and experiences could leave lasting marks on her child. This notion was often intertwined with superstitions and traditional practices related to pregnancy and childbirth.

## The Story of Disproof

The disproof of the theory of maternal impression is closely linked to the advancement of scientific understanding and the development of modern medicine. Several key factors contributed to the eventual rejection of this belief.

## William Harvey's Discoveries (17th Century)

One of the crucial turning points in the disproof of maternal impression was the work of English physician William Harvey (1578-1657). Harvey's groundbreaking discoveries regarding the circulation of blood and the process of embryogenesis laid the foundation for a more accurate understanding of fetal development. His findings suggested that the development of the fetus was driven by biological processes rather than the mother's emotions or experiences.

## Emergence of Experimental Science

As experimental science and empirical observation became more prominent in the 17th and 18th centuries, it became increasingly clear that maternal impressions could not account for the wide range of characteristics observed in children. The idea that a mother's emotions could cause physical deformities or traits in her child was inconsistent with the growing body of scientific knowledge.

## The Role of Genetics

The field of genetics, particularly the work of Gregor Mendel in the 19th century, provided further evidence against the theory of maternal impression. Mendel's experiments with pea plants demonstrated the principles of inheritance and the role of genes in determining an organism's traits. This laid the groundwork for the modern understanding of how traits are passed from parents to offspring, undermining the concept of maternal influence.

## The Decline of Maternal Impression

With the accumulation of scientific evidence, the theory of maternal impression gradually lost credibility. By the late 19th and early 20th centuries, it was widely rejected by the medical and scientific communities. Instead, the emerging fields of embryology, genetics, and developmental biology offered more comprehensive explanations for the factors influencing fetal development.

## Contemporary Understanding of Pregnancy

Today, our understanding of pregnancy is firmly grounded in biology and genetics. We recognize that a child's genetic makeup is determined by the combination of genetic material from both parents, and that external factors, such as maternal emotions or experiences, do not play a direct role in shaping physical traits or health outcomes.

## Conclusion

The theory of maternal impression is a historical belief that has its origins in ancient cultures and continued to influence perceptions of pregnancy for centuries. It posited that a pregnant woman's thoughts, experiences, or emotions could impact her unborn child's development. However, with the advancements in science, particularly in embryology, genetics, and developmental biology, the theory gradually lost credibility.

The disproof of maternal impression was a significant step toward a more accurate understanding of pregnancy and fetal development. It highlighted the importance of empirical observation, experimentation, and the role of genetics in shaping an individual's traits and characteristics. Today, our understanding of pregnancy is firmly rooted in scientific principles, dispelling the age-old notion that a mother's emotions or experiences can imprint onto her child's physical being.

# IV

# Chemistry

# Phlogiston Theory

The Phlogiston Theory was a scientific hypothesis that aimed to explain the nature of combustion and the behavior of substances during burning. It posited the existence of a hypothetical substance called "phlogiston" and was a fundamental concept in chemistry during the 18th century. In this exploration, we will delve into the nature of the theory, its origins, key proponents, the gradual decline of its credibility, and the eventual disproof of the theory.

**What Was the Phlogiston Theory?**

The Phlogiston Theory, proposed in the 17th century, was a conceptual framework used to explain the process of combustion and other chemical reactions involving the release of heat and light. It relied on the existence of a hypothetical substance called "phlogiston." According to this theory, substances that could burn contained phlogiston, which was released into the air during combustion. This release of phlogiston explained the observed phenomena of substances becoming lighter when burned and the production of ash.

**Origins and Key Proponents**

The Phlogiston Theory can be traced back to the works of several scientists and alchemists of the 17th and 18th centuries. Key figures associated with the theory include:

**Johann Joachim Becher (1635-1682):** A German alchemist and early proponent of the phlogiston concept, Becher introduced the idea of "terra pinguis," a substance that could explain the process of combustion.

**Georg Ernst Stahl (1660-1734):** Stahl, a German chemist and physician, is often credited with formalizing the Phlogiston Theory. He proposed that all combustible materials contained phlogiston, which was released during combustion.

**Joseph Priestley (1733-1804):** An English clergyman and scientist, Priestley conducted experiments related to the Phlogiston Theory. He discovered various gases, including oxygen, and interpreted his findings in terms of phlogiston.

### Scientific Reception and Challenges

The Phlogiston Theory gained widespread acceptance and became a fundamental concept in chemistry during the 18th century. However, it also faced several challenges and inconsistencies:

**Weight Gain during Combustion:** One of the fundamental problems with the theory was that it could not explain why some substances appeared to gain weight during combustion rather than losing it. For example, when metals like magnesium burned, they seemed to increase in weight.

**Discovery of Oxygen:** The discovery of oxygen by Antoine Lavoisier in the late 18th century posed a significant challenge to the Phlogiston Theory. Lavoisier demonstrated that substances burned by combining with oxygen, rather than releasing phlogiston.

**Inaccuracies in Predictions:** The Phlogiston Theory struggled to accurately predict the outcomes of chemical reactions. It could not explain why some reactions did not behave as expected based on phlogiston principles.

**Decline and Disproof**

The decline of the Phlogiston Theory can be attributed to the mounting evidence against it, particularly the discoveries related to oxygen. Antoine Lavoisier's work in the late 18th century revolutionized chemistry and led to the demise of the Phlogiston Theory. His experiments and observations demonstrated that substances combined with oxygen during combustion and that the weight of the products of combustion increased.

The disproof of the theory can be pinpointed to the late 18th century, with Lavoisier's groundbreaking work on the conservation of mass and the discovery of oxygen. In 1789, Lavoisier published "Elements of Chemistry," which provided a new and accurate framework for understanding chemical reactions and combustion. His experiments and precise measurements effectively replaced the phlogiston concept with the concept of chemical elements and compounds.

Lavoisier's work led to the establishment of modern chemistry and a more accurate understanding of chemical reactions. The Phlogiston Theory gradually lost credibility and faded into scientific history as an obsolete and inadequate explanation for combustion and chemical processes.

In summary, the Phlogiston Theory was a scientific hypothesis that proposed the existence of a hypothetical substance called phlogiston to explain combustion and chemical reactions involving the release of heat and light. It was widely accepted during the 18th century but faced challenges and inconsistencies, particularly in light of the discoveries related to oxygen. Antoine Lavoisier's work in the late 18th century, which included the discovery of oxygen and the principle of the conservation of mass, effectively disproved the Phlogiston Theory and led to the birth of modern chemistry.

# The Four Elements Theory

The Four Elements Theory, one of the foundational concepts in ancient philosophy and natural science, represents an ancient perspective on the composition of matter. In this exploration, we will uncover the theory's essence, its origins, proponents, challenges it faced, and how it eventually gave way to modern scientific understanding.

**Unraveling the Four Elements Theory**

The Four Elements Theory posits that all matter is composed of four fundamental substances, or elements, which are earth, water, air, and fire. Each element has unique properties and characteristics that determine the nature of the substances they compose.

**Origins of the Four Elements Theory**

The Four Elements Theory has ancient roots, with its foundations laid by early Greek philosophers:

**Empedocles:** Empedocles, a Greek philosopher born in the 5th century BCE, is often credited with formulating the concept of the four elements. He proposed that these elements combined and separated through the forces of love (attraction) and strife (repulsion), giving rise to all observable substances.

**Plato and Aristotle:** Both Plato and Aristotle, two of the most influential figures in ancient philosophy, further developed the Four Elements Theory. Plato believed that these elements were associated with geometric shapes, while Aristotle provided a more detailed classification of matter based on the qualities of hot, cold, wet, and dry.

**Supporters and Believers**

The Four Elements Theory held sway for centuries and found support among various prominent historical figures:

**Hippocrates:** The ancient Greek physician Hippocrates, known as the "Father of Medicine," incorporated the Four Elements Theory into his medical philosophy. He believed that an imbalance in the four humors (associated with the elements) caused illness.

**Galen:** Galen, a Roman physician, built upon the theories of Hippocrates and integrated the Four Elements Theory into his medical practice. His ideas influenced medical thought for centuries.

**Alchemy:** Alchemists in the Middle Ages adopted the Four Elements Theory as a fundamental principle in their quest to transmute base metals into gold and discover the philosopher's stone.

**The Demise of the Four Elements Theory**

The Four Elements Theory gradually began to lose its dominance as the scientific understanding of matter evolved:

**Emergence of Modern Chemistry:** The development of modern chemistry in the 17th and 18th centuries challenged the concept of the four elements. Scientists like Robert Boyle and Antoine Lavoisier conducted experiments that revealed the true nature of chemical elements and compounds.

**Discovery of the Periodic Table:** Dmitri Mendeleev's creation of the periodic table in the 19th century provided a comprehensive framework for understanding the elements based on atomic structure and properties, displacing the traditional four-element model.

**Advancements in Atomic Theory:** The atomic theory of John Dalton and subsequent developments in atomic physics demonstrated that matter is composed of individual atoms, not the four elemental substances proposed by the ancient theory.

## When the Four Elements Theory Got Disproved

The disproof of the Four Elements Theory was a gradual process spanning several centuries, with key developments occurring in the 17th and 18th centuries:

**Robert Boyle (1627-1691):** Boyle's experiments on the behavior of gases, such as his law describing the relationship between pressure and volume, challenged the ancient concept of elements. His work laid the foundation for the development of modern chemistry.

**Antoine Lavoisier (1743-1794):** Lavoisier is often regarded as the "Father of Modern Chemistry." He conducted meticulous experiments that led to the discovery of the conservation of mass and the identification of chemical elements, effectively dismantling the traditional Four Elements Theory.

**Dmitri Mendeleev (1834-1907):** Mendeleev's creation of the periodic table in the late 19th century provided a systematic classification of elements based on atomic properties and relationships, rendering the four-element model obsolete.

In conclusion, the Four Elements Theory, originating in ancient Greece and further developed by philosophers and scientists throughout history, represented an early attempt to understand the composition of matter. It posited that all substances were composed of four fundamental elements: earth, water, air, and fire. While influential for centuries, the theory gradually lost ground in the face of advancing scientific knowledge. The emergence of modern chemistry, the discovery of the periodic table, and advancements in atomic theory collectively contributed to the ultimate disproof of the Four Elements Theory, paving the way for a more accurate and comprehensive understanding of the nature of matter.

# Alchemy

Alchemy, an ancient and multifaceted discipline, was a historical precursor to modern chemistry. Rooted in a complex blend of philosophy, mysticism, and proto-scientific ideas, alchemy sought to transform base substances into noble ones, discover the philosopher's stone, and achieve the elusive goal of the Elixir of Life. The origins of alchemy date back to antiquity, and it has a rich history of practitioners and theorists, including figures like Hermes Trismegistus and Paracelsus. Despite its mystical and unscientific aspects, alchemy played a pivotal role in the development of modern chemistry and ultimately evolved into the scientific discipline we recognize today.

## The Essence of Alchemy

At its core, alchemy was concerned with the transformation of matter. Alchemists believed in the transmutation of base metals into noble ones, the discovery of the philosopher's stone (a substance that could transmute metals and grant immortality), and the pursuit of the Elixir of Life (a potion granting eternal life).

## Origins of Alchemy

The origins of alchemy are shrouded in antiquity, and its early development is difficult to pinpoint. However, the practice of manipulating and transforming metals can be traced back to ancient civilizations.

**Hermes Trismegistus:** The legendary figure Hermes Trismegistus, a combination of the Greek god Hermes and the Egyptian god Thoth, is often credited with the early development of alchemical ideas. The Hermetic texts attributed to Hermes Trismegistus laid the groundwork for alchemical philosophy.

**Chinese Alchemy:** Alchemy was independently practiced in China, where it focused on medicinal and pharmacological pursuits. Chinese alchemists sought the elixir of immortality and the transmutation of substances.

**Proponents of Alchemy**

Throughout history, alchemy attracted a diverse array of practitioners, including scientists, philosophers, and mystics.

**Paracelsus:** The Swiss physician and alchemist Paracelsus made significant contributions to alchemical theory and practice in the 16th century. He emphasized the importance of experimentation and observation in alchemical processes.

**Johann Joachim Becher:** In the 17th century, Becher proposed a theory of phlogiston, a substance central to alchemical processes, which contributed to the evolving understanding of chemical reactions.

**The Decline of Alchemy: Emergence of Modern Chemistry**

The practice of alchemy began to decline as modern science, grounded in empirical observation and experimentation, gained prominence. Several key developments contributed to the discrediting of alchemical principles.

**Scientific Revolution:** The Scientific Revolution of the 16th and 17th centuries emphasized empiricism and the development of the scientific method. Alchemical practices often lacked the systematic and empirical approaches that characterize modern science.

**Emergence of Modern Chemistry:** Alchemy played a transitional role in the development of modern chemistry. Early chemists like Robert Boyle and Antoine Lavoisier made groundbreaking discoveries that challenged alchemical ideas.

**Phlogiston Theory and Chemical Reactions:** The phlogiston theory, while an attempt to explain combustion and chemical reactions, was ultimately replaced by more accurate theories that laid the foundation for modern chemistry.

### When Alchemy Was Disproved

The decline and discrediting of alchemy occurred gradually over several centuries, culminating in the emergence of modern chemistry.

**17th Century:** The 17th century saw the rise of modern experimental science, with figures like Robert Boyle conducting controlled experiments and rejecting the mystical aspects of alchemy.

**18th Century:** The phlogiston theory, central to alchemical processes, was challenged by the work of Antoine Lavoisier and the discovery of oxygen as the key to combustion.

**19th Century:** By the 19th century, alchemy had largely given way to the emerging field of modern chemistry, which provided systematic explanations for chemical reactions and the nature of matter.

### Legacy and Insights

Alchemy's legacy is twofold. It serves as a historical precursor to modern chemistry, influencing the development of laboratory techniques, apparatus, and terminology. At the same time, alchemy's mystical and philosophical aspects continue to captivate those interested in esoteric traditions and spiritual transformation. While its practical and mystical goals remain largely unfulfilled, alchemy remains a fascinating chapter in the history of human curiosity and scientific exploration, illustrating the evolving nature of human knowledge and our enduring quest for understanding the world around us.

## The Phlogiston Theory: An Exploration of a Curious Chapter in Science

In the annals of scientific history, few theories have captured the imagination and perplexed the minds of scholars as the Phlogiston theory did. This historical narrative takes us on a journey back to a time when chemists believed that the very act of burning something was a transformative process involving a mysterious substance called "phlogiston." The tale of phlogiston theory is an intriguing one, with its roots in the late 17th century, involving notable figures like Johann Joachim Becher and Georg Ernst Stahl. While it once held sway in the world of chemistry, it ultimately met its demise in the face of groundbreaking experiments by Antoine Lavoisier.

### The Phlogiston Theory: An Enigmatic Concept

At the heart of the Phlogiston theory lies a captivating idea: that materials capable of burning, such as wood or coal, contain within them a concealed substance called phlogiston. When these materials undergo combustion, they release this hidden entity into the atmosphere, leaving behind what we now recognize as ashes. Phlogiston was believed to be the driving force behind combustion, and its release explained the fiery transformation of matter.

## A Historical Origin: The Genesis of Phlogiston

The origins of the Phlogiston theory can be traced back to a period when alchemy still played a prominent role in the scientific landscape. It was during the late 17th century that Johann Joachim Becher, a German alchemist and physician, first introduced the concept of "terra pinguis." This enigmatic and flammable substance hinted at the secrets of combustion. However, it was Becher's pupil, Georg Ernst Stahl, who refined and expanded upon this idea, shaping it into what we now know as the Phlogiston theory in the early 18th century.

## The Believers: Notable Figures Who Embraced Phlogiston

In any narrative, intriguing characters emerge, and the Phlogiston theory was no exception. Leading the charge in advocating for this theory was none other than Georg Ernst Stahl himself. In 1697, he articulated his phlogiston theories in "Zymotechnia Fundamentalis," captivating the imaginations of other scientists and scholars. Among those drawn to the allure of phlogiston were prominent figures like Johann Heinrich Pott and Johann Joachim Lange. These minds of the era found in phlogiston an elegant explanation for the transformative nature of combustion.

## The Demise of Phlogiston: A Story of Scientific Inquiry

Now, picture a narrative filled with scientific intrigue and unexpected discoveries. Enter Antoine Lavoisier, a French chemist renowned for his meticulous approach to experimentation. In the late 18th century, he set out to subject the claims of the Phlogiston theory to rigorous scrutiny, setting the stage for the theory's unraveling.

## A Revelation: The Downfall of Phlogiston

Consider this scenario: a log burning in a fireplace. According to the Phlogiston theory, as the log burns, it should lose weight because it's releasing phlogiston into the air. However, Lavoisier took an entirely different path in his experiments. He carefully measured the products of combustion and discovered something astonishing. Contrary to the predictions of the Phlogiston theory, burning substances didn't lose weight; they gained weight. It was as if they had partaken in an invisible feast during the act of combustion.

In a narrative marked by scientific curiosity, Lavoisier revealed that substances, when subjected to combustion, combined with a gas he would later dub "oxygen" (a discovery in its own right). This finding directly contradicted the central claim of the Phlogiston theory, which held that burning materials lost mass because they released phlogiston. In reality, these materials gained mass due to their interaction with oxygen.

## The Conclusion: The End of Phlogiston's Reign

With Lavoisier's groundbreaking work on the conservation of mass and the identification of oxygen, the Phlogiston theory faced its ultimate demise. By the close of the 18th century, the scientific community had collectively abandoned the notion of phlogiston in favor of Lavoisier's oxygen theory of combustion. The once-cherished idea that phlogiston was responsible for the magical transformation of matter through fire had finally been extinguished.

In summary, the Phlogiston theory, emerging in the late 17th century and cultivated by Georg Ernst Stahl, proposed the existence of phlogiston as a substance driving combustion.

Supported by scientists and scholars of the era, this theory eventually succumbed to Antoine Lavoisier's meticulous experiments, which revealed that substances gained weight during combustion due to their interaction with oxygen. The fall of phlogiston marked a significant turning point in the history of chemistry, underscoring the evolving nature of scientific understanding.

# Caloric Theory

Caloric theory was a scientific concept that dominated the understanding of heat and temperature for a significant part of scientific history. It postulated the existence of a mysterious fluid-like substance called "caloric" that was believed to be responsible for heat transfer. Originating in the 18th century, this theory gained the support of prominent scientists like Antoine Lavoisier and Pierre-Simon Laplace. However, it ultimately succumbed to the scrutiny of experimental evidence and was replaced by the kinetic theory of heat in the 19th century. In this exploration, we will delve into what caloric theory entailed, its historical origins, the early proponents who championed it, the influential figures who endorsed its validity, the gradual erosion of its credibility, and the pivotal moment when caloric theory was conclusively disproven.

## The Essence of Caloric Theory

Caloric theory was founded on the idea that heat was not a property of matter but rather a substance itself—caloric. According to this theory, caloric was an invisible, weightless fluid-like substance that could flow from one body to another, causing changes in temperature. When caloric flowed into an object, it would become hotter, and when it flowed out, it would become cooler. This concept of heat as a fluid was central to the caloric theory.

## The Birth of Caloric Theory: Origins in the 18th Century

Caloric theory emerged in the latter half of the 18th century, a period marked by significant advances in the understanding of heat and thermodynamics. One of the earliest proponents of the theory was Joseph Black, a Scottish physicist and chemist. Black's work in the 1760s on the specific heat of substances laid the groundwork for the concept of caloric as a substance that could be absorbed and released by materials.

**Antoine Lavoisier and Caloric:** The support of Antoine Lavoisier, a pioneering French chemist known for his work on the conservation of mass, gave further credence to caloric theory. Lavoisier incorporated caloric into his chemical explanations, and the theory gained prominence in the late 18th century.

## Prominent Figures Who Endorsed Caloric

Caloric theory garnered the support of several influential scientists and thinkers of its time, including Pierre-Simon Laplace, Sadi Carnot, and Lazare Carnot. These figures, renowned for their contributions to physics and thermodynamics, incorporated caloric into their work and sought to refine the theory.

## The Decline of Caloric Theory: Experimental Challenges

The eventual decline of caloric theory was precipitated by mounting experimental evidence that challenged its fundamental assumptions. One significant challenge came from the work of Count Rumford (Benjamin Thompson), an American-born British physicist, who conducted experiments in the late

18th century.

**Rumford's Experiments:** Count Rumford's experiments, notably his work on the boring of cannons in the manufacture of artillery, revealed that an immense amount of heat could be generated by mechanical work. This finding contradicted the caloric theory's central premise that heat was a substance that could be conserved. Instead, it suggested that heat could be created from mechanical motion, casting doubt on the validity of caloric as a weightless fluid.

**Carnot and the Development of Thermodynamics:** In the 19th century, Sadi Carnot's groundbreaking work on the efficiency of heat engines further challenged caloric theory. Carnot's insights laid the foundation for the modern science of thermodynamics and introduced the concept of a heat engine operating between two temperature reservoirs. His work did not require the existence of caloric as a fluid, and instead, it described heat as a form of energy.

### The Triumph of the Kinetic Theory of Heat

The final blow to caloric theory came with the emergence of the kinetic theory of heat in the mid-19th century. Scientists like James Clerk Maxwell and Ludwig Boltzmann developed this theory, which proposed that heat was the result of the motion of particles at the atomic and molecular level.

**The Role of Rudolf Clausius:** Rudolf Clausius, a German physicist, played a pivotal role in advancing the kinetic theory of heat and formalizing the concept of entropy. Clausius' work on the mathematical foundations of thermodynamics replaced caloric with the concept of energy and established the modern framework for understanding heat and temperature.

### The Disproof of Caloric Theory

Caloric theory was ultimately disproven by the accumulation of experimental evidence and the emergence of more comprehensive and predictive theories of heat and thermodynamics. By the mid-19th century, the kinetic theory of heat and the laws of thermodynamics had replaced caloric theory as the prevailing framework for understanding heat and temperature.

## Legacy and Insights

The rise and fall of caloric theory offer valuable insights into the progression of scientific knowledge. It demonstrates how deeply ingrained theories can be challenged and eventually replaced when faced with compelling experimental evidence and more accurate explanatory models. The transition from caloric theory to the kinetic theory of heat marked a transformative moment in the history of thermodynamics and paved the way for our modern understanding of heat, energy, and the behavior of matter at the atomic and molecular level.

# V

# Physics

# Aether Drag

The Aether Drag Hypothesis is a concept deeply rooted in the history of physics, representing an attempt to explain the behavior of light and its propagation through space. In this exploration, we will delve into the theory's origins, its proponents, the challenges it faced, and the ultimate path to its disproof.

### Uncovering the Aether Drag Hypothesis

The Aether Drag Hypothesis is based on the idea of a luminiferous aether—an all-pervading, invisible substance that was once thought to fill the vacuum of space. This hypothetical medium was believed to be necessary to explain the propagation of light waves, as light was considered a wave phenomenon.

### Origins of the Aether Drag Hypothesis

The Aether Drag Hypothesis was formulated in response to the growing understanding of light as a wave, particularly during the 19th and early 20th centuries:

**Augustin-Jean Fresnel:** The French physicist Augustin-Jean Fresnel made significant contributions to understanding the wave nature of light. His work on diffraction and interference patterns in light waves laid the foundation for the Aether Drag Hypothesis. Fresnel's equations suggested that light waves required a medium through which to propagate, giving rise to the concept of the luminiferous aether.

**Michelson-Morley Experiment:** The Michelson-Morley experiment of 1887 played a crucial role in the development of the Aether Drag Hypothesis. Albert A. Michelson and Edward W. Morley designed the experiment to detect the motion of Earth through the presumed aether. Surprisingly, their results showed no evidence of such motion, which posed a significant challenge to the existing theories of aether.

**Supporters and Believers**

Several prominent scientists were proponents of the Aether Drag Hypothesis, seeing it as a means to reconcile the concept of a luminiferous aether with experimental observations:

**Albert A. Michelson:** Despite the null results of the Michelson-Morley experiment, Albert A. Michelson continued to support the idea of the aether. He proposed the Aether Drag Hypothesis, suggesting that the aether was partially dragged by the motion of Earth, which explained the experiment's outcomes.

**Hendrik Lorentz:** Dutch physicist Hendrik Lorentz was a key figure in developing the Aether Drag Hypothesis. He introduced the idea of length contraction and time dilation to explain the observed null results of the Michelson-Morley experiment within the framework of aether theory.

**The Demise of the Aether Drag Hypothesis**

The Aether Drag Hypothesis faced several significant challenges and eventually lost credibility within the scientific community:

**Special Theory of Relativity:** Albert Einstein's Special Theory of Relativity, published in 1905, revolutionized the understanding of space and time. It replaced the need for a luminiferous aether with the concept of spacetime, where the speed of light is constant for all observers. This theory provided a more elegant and accurate explanation of the behavior of light and its propagation.

**General Theory of Relativity:** Einstein's General Theory of Relativity, published in 1915, further expanded upon the principles of his Special Theory. It described gravity as the curvature of spacetime rather than a force transmitted through aether. This theory gained widespread acceptance and further discredited the need for aether.

**Additional Experiments:** Subsequent experiments, such as the Kennedy-Thorndike experiment and the Trouton-Noble experiment, provided further support for the principles of relativity and undermined the Aether Drag Hypothesis.

## When the Aether Drag Hypothesis Got Disproved

The disproof of the Aether Drag Hypothesis was a gradual process, spanning the early 20th century. The publication of Albert Einstein's Special Theory of Relativity in 1905 marked a significant turning point in this regard. Einstein's theory, along with subsequent experiments that supported its predictions, effectively replaced the concept of the luminiferous aether and its associated Aether Drag Hypothesis with a more elegant and empirically validated framework.

In conclusion, the Aether Drag Hypothesis emerged as an attempt to reconcile the concept of a luminiferous aether with experimental observations related to the behavior of light. Despite the support of notable physicists like Albert A. Michelson and Hendrik Lorentz, the Aether Drag Hypothesis faced insurmountable challenges. The advent of Albert Einstein's Special Theory of Relativity in 1905, followed by his General Theory of Relativity in 1915, provided alternative explanations for the behavior of light and the nature of space and time. These theories, along with additional experiments, gradually discredited the Aether Drag Hypothesis, ultimately leading to its disproof within the scientific community.

# Aetheric Gravitation

The aetheric theory of gravitation was a scientific hypothesis that aimed to explain the force of gravity by invoking the existence of a mysterious substance called "aether." This theory originated in the late 19th century and persisted for some time, attracting the attention of notable scientists and sparking debates within the scientific community. However, it was eventually replaced by Albert Einstein's theory of general relativity in the early 20th century, which provided a more comprehensive and accurate explanation of gravitation.

**Understanding the Aetheric Theory of Gravitation**

The aetheric theory of gravitation can be summarized as follows:

**The Aether Concept:** The theory postulated the existence of a hypothetical substance called "aether" that was believed to permeate all of space. This aether was thought to be a medium through which gravitational forces were transmitted.

**Gravitational Interaction:** According to this theory, objects with mass interacted with the aether, creating gravitational fields. These fields were responsible for the attractive force of gravity that we observe on Earth.

**Aether Properties:** The properties of aether, such as its density and elasticity, were hypothesized to determine the strength and behavior of gravity.

**Origins of the Aetheric Theory of Gravitation**

The aetheric theory of gravitation emerged in the late 19th century as an attempt to reconcile the classical physics of Isaac Newton with new developments in electromagnetism:

**James Clerk Maxwell:** Maxwell's equations, which described the behavior of electromagnetic waves, had successfully explained the propagation of light and other electromagnetic phenomena. However, they raised questions about the nature of space itself and whether it was filled with a medium, like aether, to support these waves.

**George Gabriel Stokes:** Stokes, a British physicist, was among the proponents of the aetheric theory. He suggested that aether played a role not only in supporting electromagnetic waves but also in mediating gravitational interactions.

**Supporters and Believers**

The aetheric theory of gravitation attracted some prominent scientists and thinkers of the time:

**Oliver Lodge:** The British physicist Sir Oliver Lodge was a proponent of the aether theory and believed that aether was the medium responsible for both electromagnetic and gravitational interactions.

**William Thomson (Lord Kelvin):** Thomson, a renowned physicist and engineer, was initially open to the idea of aether as a medium for gravity but later became skeptical as experimental evidence failed to support the theory.

**The Demise of the Aetheric Theory of Gravitation**

The aetheric theory of gravitation faced several challenges and eventually gave way to Albert Einstein's theory of general relativity:

**Michelson-Morley Experiment:** In 1887, the Michelson-

Morley experiment attempted to detect the motion of the Earth through the aether by measuring the speed of light in different directions. The results showed no evidence of an aether, casting doubt on its existence as a medium for light or gravity.

**Einstein's Theory of General Relativity:** In 1915, Albert Einstein introduced the theory of general relativity, which revolutionized our understanding of gravitation. Instead of relying on an aetheric medium, Einstein's theory described gravity as the curvature of spacetime caused by mass and energy. General relativity not only explained classical gravitational phenomena but also made accurate predictions, such as the bending of light by gravity.

**When the Aetheric Theory of Gravitation Got Disproved**

The aetheric theory of gravitation gradually lost credibility and was largely replaced by the advent of general relativity:

**Late 19th and Early 20th Century:** The theory gained attention and support from some scientists during this period, but the lack of experimental evidence in its favor, combined with the success of other theories like special and general relativity, led to its decline.

**1915:** Albert Einstein published his theory of general relativity, which provided a new framework for understanding gravitation. This marked the beginning of the end for the aetheric theory of gravitation.

In conclusion, the aetheric theory of gravitation was a hypothesis that proposed the existence of aether as a medium for transmitting gravitational forces. It emerged in response to developments in electromagnetic theory but faced challenges from experiments like the Michelson-Morley experiment and was ultimately replaced by Albert Einstein's theory of general relativity. Despite the enthusiasm of some scientists and proponents, the aetheric theory of gravitation could not withstand the weight of evidence against it, paving the way for a more accurate understanding of gravity in the modern era.

# Luminiferous Aether

Luminiferous aether was a hypothetical medium believed to permeate all of space and serve as the medium through which light waves propagated. This theory, which originated in the 19th century, aimed to provide an explanation for how light, a wave phenomenon, could travel through the seemingly empty vacuum of space. Luminiferous aether was proposed by scientists and philosophers of the time as an attempt to reconcile the wave nature of light with the apparent absence of a material medium. It was an influential concept in the history of physics and was supported by some of the most prominent scientists of the era. However, as the understanding of light and the universe evolved, the theory of luminiferous aether gradually lost favor and was eventually replaced by Albert Einstein's theory of special relativity. In this exploration, we will delve into the essence of luminiferous aether, its historical origins, its early proponents, the influential individuals who believed in its validity, the process by which it was ultimately disproved, and the timeline of its discrediting.

**The Essence of Luminiferous Aether**

Luminiferous aether was proposed as a hypothetical medium that filled all of space and served as the medium through which light waves were believed to travel. It was conceived to explain how light, a wave phenomenon, could propagate through a seemingly empty vacuum.

**Origins of Luminiferous Aether**

The origins of the concept of luminiferous aether can be traced to the 19th century, a period marked by significant developments in the understanding of light and electromagnetic phenomena.

**Thomas Young and Wave Theory:** The wave theory of light, championed by scientists like Thomas Young, gained prominence in the early 19th century. Young's double-slit experiment demonstrated the interference pattern of light waves, supporting the wave nature of light.

**Augustin-Jean Fresnel's Contributions:** Augustin-Jean Fresnel, a French physicist, made significant contributions to the wave theory of light. His work on diffraction and polarization provided theoretical foundations for the concept of luminiferous aether.

**Proponents of Luminiferous Aether**

Luminiferous aether had influential proponents within the scientific community of the 19th century, including Fresnel and others.

**Fresnel's Wave Theory:** Fresnel's mathematical model for the behavior of light waves incorporated the idea of luminiferous aether as the medium through which light propagated.

**Éleuthère Mascart's Experiments:** French physicist Éleuthère Mascart conducted experiments in the mid-19th century to detect the motion of the Earth through the supposed aether. While these experiments did not yield conclusive results, they were indicative of the prevailing belief in the existence of the aether.

## The Decline of Luminiferous Aether: Emergence of Special Relativity

The credibility of luminiferous aether began to wane as the late 19th and early 20th centuries witnessed advancements in the understanding of light and the theory of relativity.

**Michelson-Morley Experiment (1887):** Albert A. Michelson and Edward W. Morley conducted a landmark experiment to detect the motion of the Earth through the aether. Their null result, which showed no evidence of aether wind, challenged the concept of luminiferous aether.

**Special Theory of Relativity (1905):** Albert Einstein's theory of special relativity, published in 1905, introduced the concept that the speed of light in a vacuum is constant for all observers. This groundbreaking theory rendered the need for a luminiferous aether obsolete.

### When Luminiferous Aether Was Disproved

The gradual decline and discrediting of the concept of luminiferous aether took place over several decades, culminating in the acceptance of Einstein's special theory of relativity.

**1887:** The Michelson-Morley experiment provided strong evidence against the existence of luminiferous aether by showing a null result for the aether wind.

**Early 20th Century:** Albert Einstein's theory of special relativity, published in 1905, introduced the principle of the constancy of the speed of light and effectively rendered the concept of luminiferous aether unnecessary.

### Legacy and Insights

Luminiferous aether represents a pivotal chapter in the history of physics. While it was once a widely accepted concept used to explain the behavior of light, it was ultimately unable to withstand experimental scrutiny. Its decline and replacement by the theory of special relativity exemplify the dynamic nature of scientific progress and the importance of empirical evidence in shaping our understanding of the universe. Luminiferous aether's legacy lies in its role as a stepping stone in the evolution of modern physics, particularly in the transition from classical to relativistic physics. It serves as a reminder that even well-established scientific theories can be challenged and replaced by more accurate models when new evidence and insights emerge.

## Ether Theory: The Elusive Substance of Cosmic Space

Ether theory was a conceptual framework that sought to explain the propagation of light and electromagnetic waves through space by postulating the existence of a luminiferous ether—a mysterious, all-pervading medium filling the cosmos. For a significant portion of scientific history, this theory held sway, with luminaries like James Clerk Maxwell and Lord Kelvin embracing it. However, it eventually succumbed to the transformative advances of the early 20th century, particularly Albert Einstein's theory of special relativity. In this exploration, we will delve into what ether theory entailed, when it first emerged, the identity of its early proponents, the influential figures who supported it, the story of its downfall, and the historical moment when ether theory was conclusively disproven.

### The Essence of Ether Theory

Ether theory was a bold proposal aimed at explaining

the propagation of electromagnetic waves, such as light, in the seemingly empty expanse of space. It postulated the existence of a luminiferous ether—a medium that permeated the universe and served as the conduit for these waves. Ether was envisioned as a substance with unique properties that allowed it to transmit electromagnetic vibrations.

### Origins and Early Notions of Ether

The roots of ether theory can be traced back to the 17th century, when early scientists grappled with the nature of light. One of the earliest proponents of the idea of an ether-like substance was René Descartes, who suggested the existence of a "plenum," a substance filling space that allowed for the transmission of light waves. However, it was the 18th and 19th centuries that witnessed significant developments in ether theory.

**Thomas Young and Wave Theory:** Thomas Young, an English scientist, played a pivotal role in advancing the concept of ether. In the early 19th century, Young proposed the wave theory of light, suggesting that light was a form of wave motion that required a medium for propagation. Although he did not specify the properties of the medium, his ideas laid the groundwork for later formulations of ether theory.

**James Clerk Maxwell and Electromagnetism:** The eminent Scottish physicist James Clerk Maxwell made significant contributions to the development of ether theory in the mid-19th century. Maxwell's equations, which describe the behavior of electric and magnetic fields, seemed to require a medium through which electromagnetic waves could travel. He posited the existence of a subtle and elastic ether to carry these waves.

**Lord Kelvin's Influence:** Another prominent figure in the promotion of ether theory was Lord Kelvin (William Thomson), a celebrated physicist and engineer. Kelvin's work on the mathematical description of heat and his contributions to the understanding of electricity and magnetism reinforced the notion of ether as the medium for electromagnetic waves.

**The Downfall of Ether Theory: Einstein's Special Relativity**

The turning point in the history of ether theory came with Albert Einstein's theory of special relativity, presented in 1905. Special relativity introduced a fundamental shift in the understanding of space and time and had profound implications for the concept of ether.

**Einstein's Postulates:** In his theory, Einstein presented two postulates that revolutionized physics. The first postulate stated that the laws of physics are the same for all observers in uniform motion, challenging the notion of an absolute reference frame tied to ether. The second postulate posited that the speed of light in a vacuum is constant for all observers, regardless of their motion.

**The Michelson-Morley Experiment:** Einstein's theory gained support from the famous Michelson-Morley experiment conducted in 1887 by Albert A. Michelson and Edward W. Morley. The experiment was designed to detect the motion of Earth through the ether by measuring the speed of light in different directions. However, the results consistently showed no evidence of an ether wind, contradicting the predictions of ether theory.

**Einstein's Triumph:** Einstein's theory of special relativity reconciled the experimental findings of Michelson and Morley with the behavior of light and electromagnetic waves.

It dispensed with the need for an ether-like medium and introduced a revolutionary perspective on space, time, and motion. According to special relativity, space and time were interwoven into spacetime, and the speed of light was an absolute constant.

## The End of Ether Theory

With the acceptance of Einstein's special relativity, ether theory gradually lost its scientific credibility. The notion of a luminiferous ether as the medium for electromagnetic waves faded into the annals of scientific history. By the early 20th century, ether theory had been effectively superseded by the new theoretical framework of special relativity.

## Legacy and Lessons

Ether theory serves as a reminder of the dynamic nature of scientific knowledge and the importance of empirical evidence in shaping our understanding of the universe. While it once held a prominent place in the scientific canon, it ultimately yielded to a more comprehensive and experimentally validated theory. The downfall of ether theory marked a transformative moment in the history of physics, paving the way for the development of modern theories that continue to shape our understanding of the cosmos.

# N-Rays

The N-Ray Theory, a curious and short-lived episode in the history of physics, emerged at the turn of the 20th century, poised to challenge established notions about the nature of radiation. Named after the University of Nancy, this theory postulated the existence of a new type of radiation, the enigmatic "N-rays." Despite initial excitement and the endorsement of some prominent scientists, the N-Ray Theory ultimately crumbled under the weight of skepticism and rigorous scientific scrutiny.

The N-Ray Theory, like many scientific theories, had its origins in the pursuit of understanding the mysteries of the natural world. It emerged in the year 1903 when French physicist René Blondlot began conducting experiments to investigate a peculiar phenomenon he believed he had discovered. Blondlot, a respected scientist and professor at the University of Nancy, was an accomplished experimentalist known for his work in optics and electromagnetism. He had gained a reputation for meticulous experimentation and was highly regarded in the scientific community.

Blondlot's interest was piqued when he observed what he believed to be a new form of radiation emitted by certain materials when exposed to X-rays. This radiation, he named

"N-rays," with "N" presumably standing for "Nancy," his university's location. According to Blondlot, N-rays possessed unique properties, including the ability to affect various materials in ways distinct from X-rays or other known forms of radiation. In particular, he claimed that N-rays could alter the brightness of phosphorescent screens and affect the index of refraction of certain substances.

Blondlot's findings quickly garnered attention and excitement, both within and outside the scientific community. At the time, radiation research was a rapidly evolving field, with discoveries like X-rays and radioactivity reshaping our understanding of the physical world. Consequently, the idea of a new type of radiation was met with considerable interest and enthusiasm.

One significant factor contributing to the credibility of the N-Ray Theory was Blondlot's own reputation as a reputable scientist. Moreover, other prominent figures in the scientific community, including renowned physicists like Henri Poincaré and Paul Langevin, expressed their support for Blondlot's work. Their endorsements lent a degree of legitimacy to the theory, further fueling its acceptance within scientific circles.

The story of the N-Ray Theory's eventual downfall is a testament to the self-correcting nature of science. While initial experiments seemed to confirm the existence of N-rays, the situation changed when scientists outside Blondlot's circle began to scrutinize his work more closely. One critical turning point occurred when American physicist Robert Wood visited Blondlot's laboratory in Nancy in 1904. Wood conducted his own experiments and, to his surprise, failed to detect any N-rays.

Skepticism grew as other scientists attempted to replicate Blondlot's experiments but encountered similar difficulties in detecting N-rays. The scientific community's faith in the theory began to wane as doubts mounted. In addition to Wood, other influential physicists, including Lord Rayleigh in the UK, expressed skepticism and questioned the existence of N-rays.

The most famous incident that led to the demise of the N-Ray Theory took place during a visit to Blondlot's laboratory by the renowned physicist and skeptic, Robert W. Wood. Wood suspected that there might be errors in Blondlot's experiments, potentially caused by psychological factors or unintentional bias. To test this hypothesis, he devised a clever plan.

During one of Blondlot's experiments, Wood secretly removed a critical component that supposedly detected N-rays. Remarkably, Blondlot continued to confidently report observations of N-rays, unaware that the key apparatus had been removed. This experiment, known as the "invisible ray" incident, demonstrated that Blondlot and his collaborators were, in essence, "seeing" N-rays because they expected to see them, not because they genuinely existed.

The "invisible ray" episode dealt a severe blow to the credibility of N-rays and Blondlot's research. It became apparent that the phenomenon was not the result of a new form of radiation but rather a case of scientific enthusiasm leading to self-deception. Subsequent investigations and criticisms further exposed flaws in Blondlot's experiments and methodology.

As a result of mounting evidence against the theory, coupled with public ridicule and skepticism within the scientific community, the N-Ray Theory began to crumble. Gradually, scientists distanced themselves from the theory, and its credibility evaporated. René Blondlot, once a respected physicist, became the subject of widespread ridicule and criticism. His work on N-rays came to be seen as an embarrassing episode in the history of science.

In 1905, after several years of controversy and skepticism, the N-Ray Theory was effectively discredited, and it faded into obscurity. The scientific consensus had shifted decisively against the existence of N-rays, and the theory became a cautionary tale of the pitfalls of confirmation bias and the importance of rigorous experimentation in the scientific process.

In conclusion, the N-Ray Theory, originating in 1903, was a short-lived but fascinating chapter in the history of physics. Named after its discoverer, René Blondlot, this theory proposed the existence of a new form of radiation called "N-rays." Blondlot, a respected physicist, garnered initial support from prominent scientists, but the theory ultimately collapsed under the weight of skepticism and critical examination.

The turning point in debunking the N-Ray Theory was the famous "invisible ray" incident, where physicist Robert Wood exposed the influence of expectation and psychological factors in Blondlot's experiments. This revelation, coupled with failed attempts to replicate Blondlot's results, led to a loss of credibility for the theory. The scientific community distanced itself from N-rays, and Blondlot's once-promising career suffered a significant setback.

The N-Ray Theory serves as a cautionary tale about the importance of skepticism, rigorous experimentation, and independent verification in the scientific process. While it briefly captured the imagination of the scientific world, it ultimately crumbled when subjected to scrutiny, illustrating the self-correcting nature of science and the need for unwavering commitment to the pursuit of truth.

# Cosmological Stagnation

The Stagnation Theory, a speculative concept in cosmology, emerged in the early 20th century as an alternative model for the universe's behavior. Originating during a time when our understanding of the cosmos was still in its infancy, this theory posited that the universe was in a state of perpetual stasis, devoid of expansion or contraction. While it may seem counterintuitive in light of modern cosmological knowledge, the Stagnation Theory held sway for a brief period and garnered attention from some notable scientists before its eventual demise.

At its core, the Stagnation Theory proposed that the universe, instead of experiencing the expansion and evolution described by the Big Bang theory, remained static and unchanging throughout eternity. According to this model, matter and energy were in a delicate equilibrium, with no net change over time. In essence, it suggested that the universe had always existed in its current form and would continue to do so indefinitely.

The origins of the Stagnation Theory can be traced back to the early 20th century, a time when our understanding of the cosmos was still relatively limited compared to the sophisticated models and observational techniques of today.

It was an era when cosmology was in its infancy, and many competing theories vied for acceptance.

One of the key proponents of the Stagnation Theory was the British astronomer and cosmologist Edward Arthur Milne. In 1935, Milne published a paper titled "A Newtonian Expanding Universe" in the journal Nature, outlining his ideas about a static, non-expanding universe. His work was based on the mathematical formalism of Albert Einstein's theory of general relativity, but Milne proposed a different interpretation of its implications.

At the time, there was still significant uncertainty surrounding the nature of the universe and its expansion. Einstein himself had introduced a "cosmological constant" into his equations to maintain a static universe, though he later abandoned it when evidence of cosmic expansion emerged.

The Stagnation Theory gained traction among a subset of scientists who were intrigued by Milne's mathematical model and its potential implications. It appealed to those who were uncomfortable with the notion of an expanding universe or who sought alternatives to the Big Bang model.

However, the demise of the Stagnation Theory began with the accumulation of observational evidence that contradicted its central premise. As technology advanced and astronomical observations became more precise, it became increasingly clear that the universe was not static but expanding.

One of the pivotal pieces of evidence came from the work of American astronomer Edwin Hubble. In the 1920s, Hubble observed distant galaxies and found that their light exhibited redshift, indicating that they were moving away from us. This discovery, known as Hubble's Law, provided compelling evidence for the expansion of the universe.

Hubble's observations and subsequent research by other astronomers established that galaxies were receding from one another, implying that the universe was expanding over time. The concept of a static universe, as proposed by the Stagnation Theory, was incompatible with this observational evidence.

The Stagnation Theory's ultimate downfall came as the weight of observational data supporting an expanding universe became overwhelming. In 1929, Hubble's landmark discovery was complemented by observations of cosmic microwave background radiation, a faint remnant of the Big Bang itself. This radiation, discovered in 1964 by Arno Penzias and Robert Wilson, provided further confirmation of the universe's dynamic and evolving nature.

In light of this growing body of evidence, the Stagnation Theory was gradually abandoned by the scientific community. While it had garnered attention and some support during its brief period of prominence, the overwhelming weight of observational data, theoretical developments, and consensus within the field led to its rejection.

The Stagnation Theory was officially debunked in the mid-20th century when the Big Bang theory, with its expanding universe and cosmic microwave background radiation, became the prevailing cosmological model. This transition marked a significant turning point in our understanding of the universe's origins and evolution.

In summary, the Stagnation Theory, which proposed a static and unchanging universe, emerged in the early 20th century as an alternative to the expanding universe model. Edward Arthur Milne, a British astronomer, was one of its key proponents. However, the theory's demise began with the accumulation of observational evidence, particularly Edwin Hubble's discovery of an expanding universe and the detection of cosmic microwave background radiation. As these lines of evidence grew stronger, the Stagnation Theory lost support within the scientific community and was ultimately discarded. This episode serves as a reminder of the ever-evolving nature of scientific knowledge and the importance of empirical evidence in shaping our understanding of the cosmos.

# The Dual-Fluid Model

The Dual-Fluid Model, an intriguing yet ultimately debunked theory in the field of physics, was born out of the desire to understand the fundamental nature of matter and its role in the universe. This model, which posited the existence of two distinct fluids, one responsible for gravitational attraction and the other for electrical interactions, represents a historical episode in the evolution of scientific thought. Though it once attracted the attention of prominent scientists, the Dual-Fluid Model eventually succumbed to the weight of evidence and the emergence of more accurate theories.

At its core, the Dual-Fluid Model proposed the existence of two fundamental fluids that permeate the universe. These fluids were postulated to be responsible for the fundamental forces of gravity and electromagnetism. In this model, one fluid was associated with gravitational attraction, while the other was linked to electrical interactions, including both electrical repulsion and attraction.

The origins of the Dual-Fluid Model can be traced back to the late 18th and early 19th centuries when scientists were beginning to grapple with the nature of forces and the fundamental constituents of matter. The model was not the work of a single individual but rather emerged from the collective musings of several scientists of the time.

One of the early proponents of this concept was Charles-Augustin de Coulomb, a French physicist renowned for his work in electrostatics. Coulomb's experiments with electrical charges led him to propose a "fluid of electricity" that permeated space and mediated electrical interactions. This fluidic notion was consistent with the prevailing ideas of the time, which often invoked invisible fluids to explain physical phenomena.

Another influential figure in the development of the Dual-Fluid Model was Charles-François de Cisternay du Fay, a French physicist and contemporary of Coulomb. Du Fay conducted experiments on electricity and proposed the existence of two electric fluids: vitreous (positive) and resinous (negative). This early formulation hinted at the idea that forces could be mediated by distinct fluids.

The concept of dual fluids gained further prominence in the early 19th century with the work of Augustin-Jean Fresnel, a French physicist known for his contributions to optics. While investigating the wave nature of light, Fresnel introduced the idea of an "ether" that pervaded space and served as the medium through which light waves propagated. Although not precisely the same as the fluids in the Dual-Fluid Model, the notion of an ether or medium underlying physical interactions was consistent with the general concept.

The Dual-Fluid Model gained traction and acceptance among scientists of the era, as it appeared to offer a unifying framework for understanding both gravitational and electrical phenomena. It provided an alternative to the previous idea of action at a distance, suggesting that forces were mediated by these mysterious fluids.

The downfall of the Dual-Fluid Model was rooted in the emergence of new experimental evidence and theoretical developments. In particular, the advent of James Clerk Maxwell's equations of electromagnetism in the mid-19th century marked a profound shift in the understanding of electrical and magnetic interactions.

Maxwell's equations, which described how electric and magnetic fields propagate through space, provided a unified and mathematically elegant framework for understanding electromagnetism. They dispensed with the need for electric fluids and demonstrated that electric and magnetic fields were interconnected, paving the way for the unification of electricity and magnetism into a single theory.

Another significant blow to the Dual-Fluid Model came from the field of gravitational physics. Albert Einstein's theory of general relativity, published in 1915, offered a revolutionary explanation for gravitation. Instead of invoking a gravitational fluid or field, Einstein's theory posited that gravity was the result of the curvature of spacetime by massive objects. This new framework provided a more accurate description of gravitational phenomena and supplanted the need for a gravitational fluid.

By the early 20th century, the Dual-Fluid Model had lost its standing as a viable explanation for fundamental forces. The emergence of Maxwell's equations, Einstein's theory of general relativity, and subsequent developments in theoretical physics provided a more robust and coherent framework for understanding the universe's fundamental forces.

In summary, the Dual-Fluid Model, which proposed the existence of two distinct fluids responsible for gravitational and electrical interactions, emerged in the late 18th and early 19th centuries.  Prominent figures like Coulomb, du Fay, and Fresnel contributed to its development, and it gained acceptance in scientific circles of the time. However, it gradually lost favor as new experimental evidence and theoretical advancements, such as Maxwell's equations and Einstein's theory of general relativity, provided more accurate and comprehensive explanations for the fundamental forces of the universe. The Dual-Fluid Model serves as a historical reminder of the evolving nature of scientific theories and the importance of empirical evidence and theoretical rigor in shaping our understanding of the physical world.

# Aristotle's Theory of Motion

Aristotle's theory of motion, an influential but ultimately flawed conception of how objects move, represents a significant chapter in the history of physics and natural philosophy. Originating in ancient Greece, this theory held sway for centuries and was embraced by numerous scholars, including Aristotle himself. However, it eventually fell out of favor as more accurate theories of motion emerged during the Scientific Revolution, marking a profound shift in our understanding of the physical world.

Aristotle's theory of motion is deeply rooted in his philosophical and scientific works, particularly his treatise "Physics" (also known as "Physica" or "Natural Philosophy"). According to Aristotle, objects are composed of four fundamental elements: earth, water, air, and fire, with each element having its natural place in the cosmos. Earth and water, being heavy elements, have a natural tendency to move downward toward the center of the universe, while air and fire, as lighter elements, have a natural tendency to move upward.

Aristotle's theory of motion is intricately linked to his concept of "natural motion" and "violent motion." Natural motion, as he described it, is the inherent, unforced motion that an object exhibits due to its elemental composition and its pursuit of its natural place in the cosmos. For instance, a stone falls downward because it consists primarily of earth and water and seeks to reach its natural place, which is the center of the Earth. Similarly, smoke rises because it is primarily composed of air and fire.

In contrast, violent motion is external or forced motion, imposed on an object by an external agent. According to Aristotle, an object would come to rest once the external force was removed, returning to its natural state of rest or natural motion.

Aristotle's theory of motion, with its emphasis on natural places and tendencies, was deeply rooted in his philosophical framework. It provided an explanation for why objects moved and the direction of their motion based on their elemental composition.

Aristotle's ideas about motion and the physical world held considerable sway for centuries, influencing not only natural philosophy but also theology and everyday understanding. His works were preserved and disseminated in various cultures and languages, contributing to their enduring influence.

Aristotle's theory of motion began to unravel with the advent of the Scientific Revolution in the 16th and 17th centuries. One of the pivotal figures in this transformation was Galileo Galilei, an Italian physicist and astronomer. Galileo conducted groundbreaking experiments on motion and gravity, challenging many of Aristotle's ideas.

Galileo's experiments with inclined planes and falling objects provided empirical evidence that contradicted Aristotle's theory of motion. He demonstrated that objects of different masses fell at the same rate in a vacuum, refuting the notion that the rate of fall was determined by an object's elemental composition or natural tendencies. This experiment led to the formulation of the principle of inertia, which states that objects in motion tend to stay in motion unless acted upon by an external force.

Furthermore, Galileo's observations of celestial bodies, particularly his telescopic observations of the Moon, Jupiter's moons, and the phases of Venus, cast doubt on the geocentric model of the universe, which was closely tied to Aristotle's ideas about the natural order of the cosmos. These observations provided strong evidence in favor of the heliocentric model proposed by Nicolaus Copernicus, which eventually replaced the geocentric model.

The downfall of Aristotle's theory of motion and his broader cosmological framework continued with the work of Sir Isaac Newton. In his monumental work "Philosophiæ Naturalis Principia Mathematica" (Mathematical Principles of Natural Philosophy), published in 1687, Newton laid the foundation for classical mechanics. He introduced his three laws of motion, which provided a comprehensive and mathematically rigorous framework for understanding the motion of objects.

Newton's laws of motion supplanted Aristotle's ideas about motion by providing a precise mathematical description of how objects move under the influence of forces. These laws, along with his law of universal gravitation, explained not only terrestrial motion but also the celestial mechanics of planets and moons. They represented a fundamental departure from Aristotle's qualitative and philosophical approach to motion, ushering in a new era of scientific inquiry characterized by quantitative analysis and empirical experimentation.

Aristotle's theory of motion was effectively disproved by the overwhelming empirical evidence and the mathematical rigor of Newtonian mechanics. Newton's laws of motion and law of universal gravitation became the cornerstone of classical physics and remained unchallenged until the advent of Albert Einstein's theory of relativity in the early 20th century.

In conclusion, Aristotle's theory of motion, which postulated natural and violent motion based on elemental composition and natural tendencies, was an influential but ultimately flawed conception of how objects move. Rooted in ancient Greek philosophy, it held sway for centuries until it was challenged by empirical evidence and more accurate theories of motion during the Scientific Revolution. Galileo's experiments and Newton's laws of motion provided a solid foundation for understanding motion and replaced Aristotle's qualitative and philosophical framework with a precise mathematical description. Aristotle's theory of motion represents a historical transition from ancient philosophical thought to the empirical and mathematical rigor of modern physics.

# Euler's Spherical Gravitation Theory

Euler's Spherical Gravitation Theory, an intriguing but ultimately refuted hypothesis in the realm of physics and astronomy, offers a fascinating glimpse into the evolving understanding of gravitational forces during the 18th century. Originating in the intellectual milieu of the Enlightenment, this theory proposed an unconventional concept of gravity as a force exerted by a surrounding spherical shell of matter. While it garnered some attention from prominent scientists of the time, Euler's theory eventually succumbed to mathematical inconsistencies and empirical contradictions.

At its core, Euler's Spherical Gravitation Theory posited that the force of gravity arises from the cumulative effect of matter distributed throughout a spherical shell surrounding an object, rather than the more familiar concept of gravity as a force originating from a point source at the center of mass. In other words, Euler suggested that the Earth's gravitational pull on an object was due to the gravitational contributions of all the matter in the celestial sphere surrounding that object.

Euler's theory was an attempt to grapple with the complexities of gravitational forces within the framework of Newtonian physics, which described gravity as a force of attraction between two point masses, governed by the inverse square law. This prevailing Newtonian view had successfully explained a wide range of celestial phenomena, from planetary motion to the orbits of moons and comets.

The origin of Euler's Spherical Gravitation Theory can be traced back to the 18th century, a period marked by significant intellectual and scientific developments. Leonhard Euler, the Swiss mathematician and physicist after whom the theory is named, was one of the most prolific and influential mathematicians of his time. Euler made substantial contributions to a variety of fields, including mechanics, number theory, and celestial mechanics.

Euler's proposal of the theory in question occurred during a period when the nature of gravitation and celestial mechanics were subjects of intense study and debate among scientists and philosophers. The Enlightenment era fostered a climate of intellectual exploration and critical thinking, which led to challenges and revisions of existing scientific paradigms.

Euler's theory attracted some attention and endorsement from notable scientists of the era, including Jean le Rond d'Alembert, a prominent mathematician and co-editor of the "Encyclopédie," a comprehensive reference work that aimed to summarize the knowledge of the time. D'Alembert believed that Euler's theory offered a novel perspective on gravity and its consequences.

However, the theory faced a number of challenges and inconsistencies that would eventually lead to its discrediting. One of the key issues was the mathematical difficulties associated with calculating the gravitational effects of a spherical shell of matter surrounding an object. While Euler made progress in developing mathematical equations to describe these effects, the calculations became exceedingly complex, and their results did not align with observed celestial phenomena.

Another significant problem with Euler's theory arose from its predictions regarding the Earth's equatorial bulge. According to his theory, the gravitational forces acting on an object at the equator should result in a significant flattening of the Earth. This contradicted empirical measurements of the Earth's shape, which indicated a near-spherical shape with a slight bulge at the equator. Euler's theory failed to provide an accurate explanation for this observed phenomenon.

The most significant blow to Euler's Spherical Gravitation Theory came from the works of Pierre-Simon Laplace and others who extended the mathematical rigor of celestial mechanics. Laplace's investigations into the dynamics of celestial bodies led to the development of potential theory and the concept of gravitational potential, which offered a more coherent and accurate framework for understanding gravity. Laplace's mathematical formulations allowed for precise calculations of gravitational effects and better explained observed phenomena.

As Laplace's work gained prominence and the mathematical inconsistencies of Euler's theory became apparent, Euler's Spherical Gravitation Theory gradually fell out of favor within the scientific community. Laplace's mathematical rigor and the success of his celestial mechanics framework contributed to the eventual rejection of Euler's theory.

In summary, Euler's Spherical Gravitation Theory, proposed by the eminent mathematician Leonhard Euler during the Enlightenment era, introduced a novel but ultimately untenable concept of gravity as a force arising from the cumulative effects of matter within a surrounding spherical shell. Despite some initial support from notable scientists of the time, including d'Alembert, the theory faced mathematical complexities and empirical contradictions that rendered it unsustainable. The advent of Laplace's potential theory and the success of his celestial mechanics framework marked the decline of Euler's theory and its eventual discrediting within the scientific community. Euler's theory serves as a historical footnote in the development of gravitational theory, highlighting the evolving understanding of gravity during a period of intellectual exploration and scientific progress.

# The Ultraviolet Catastrophe

The Ultraviolet Catastrophe, a significant problem in the history of physics, emerged in the late 19th century as a perplexing puzzle that challenged classical physics and the understanding of black-body radiation. This theoretical conundrum, rooted in the science of electromagnetism, thermodynamics, and statistical mechanics, shed light on the limitations of classical physics and played a pivotal role in the development of quantum theory. The Ultraviolet Catastrophe, while never believed to be a genuine catastrophe, spurred a profound transformation in the world of physics.

At its core, the Ultraviolet Catastrophe was a problem related to the distribution of energy in the electromagnetic spectrum, specifically concerning the radiation emitted by a black body. A black body is an idealized object that absorbs all incident electromagnetic radiation and emits radiation across a range of frequencies, depending on its temperature. The problem was to determine the spectral distribution of this emitted radiation.

The theory originated in the late 19th century when physicists were actively exploring the behavior of electromagnetic radiation and its connection to temperature. One of the pioneers in this field was Max Planck, a German physicist who made groundbreaking contributions to the understanding of radiation and black-body spectra.

In the late 19th century, classical physics was well-established, and scientists believed they could accurately describe the behavior of physical systems using classical mechanics and electromagnetism. However, when it came to explaining the spectrum of radiation emitted by a black body, classical physics encountered a profound and unexpected challenge.

The classical approach to understanding black-body radiation, based on electromagnetic theory and statistical mechanics, led to a prediction that contradicted experimental observations. According to classical physics, as the temperature of a black body increased, the intensity of radiation at all frequencies should also increase without bound. In other words, the classical theory implied that the ultraviolet radiation emitted by a hot black body should be infinitely intense, a prediction that became known as the "Ultraviolet Catastrophe."

This prediction was in stark contrast to experimental data, which showed that the intensity of black-body radiation increased with temperature but did not exhibit an infinite and unbounded growth in the ultraviolet region. This glaring discrepancy between theory and experiment raised a fundamental question about the limitations of classical physics in describing the behavior of physical systems at the atomic and molecular level.

The Ultraviolet Catastrophe posed a significant challenge to the leading physicists of the time, including Lord Rayleigh and Sir James Jeans. These scientists attempted to reconcile the classical predictions with experimental results by introducing ad hoc modifications to the theory, such as the "Rayleigh-Jeans law." However, these modifications were unsatisfactory and could not resolve the fundamental issues raised by the catastrophic predictions.

The breakthrough in resolving the Ultraviolet Catastrophe came from Max Planck, who recognized that the classical approach to understanding black-body radiation was fundamentally flawed. Planck introduced a radical departure from classical physics by proposing that energy is quantized, meaning it can only exist in discrete packets or quanta. He postulated that the energy of electromagnetic radiation was quantized in discrete units proportional to its frequency, with a constant of proportionality now known as Planck's constant (h).

Planck's quantum hypothesis was revolutionary. It implied that energy levels were not continuous, as classical physics had assumed, but discrete and quantized. This insight allowed Planck to derive a formula for black-body radiation, known as Planck's radiation law, which perfectly matched the experimental data and resolved the Ultraviolet Catastrophe.

Planck's work on black-body radiation laid the foundation for quantum mechanics, a new branch of physics that fundamentally transformed our understanding of the microscopic world. His quantum hypothesis introduced the concept of quantization of energy and initiated a paradigm shift that led to the development of quantum theory by subsequent physicists, including Albert Einstein, Niels Bohr, and Erwin Schrödinger.

The Ultraviolet Catastrophe was effectively disproved by Planck's quantum theory, which provided a robust and accurate description of black-body radiation. Planck's revolutionary ideas and mathematical formulations not only resolved the problem but also laid the groundwork for the quantum revolution that would shape the future of physics.

In summary, the Ultraviolet Catastrophe was a theoretical problem in the late 19th century related to the distribution of energy in the electromagnetic spectrum emitted by a black body. It challenged classical physics, as it predicted unbounded and infinitely intense radiation in the ultraviolet region, contrary to experimental observations. Max Planck's introduction of the quantum hypothesis, postulating the quantization of energy, resolved the problem and led to the development of quantum mechanics. The Ultraviolet Catastrophe played a crucial role in the transformation of physics and our understanding of the fundamental nature of matter and energy.

# Cosmic Rays and Evolution

The notion that cosmic rays could influence evolution is a historical scientific hypothesis that emerged during the 20th century. Proposed as a speculative idea, it postulated that cosmic rays, which are high-energy particles originating in space, might play a role in inducing genetic mutations and thereby driving evolutionary processes on Earth. While this theory initially gained some attention and even support from notable scientists, it ultimately faced significant challenges and was largely disproved as our understanding of radiation biology and genetics evolved.

The theory of cosmic rays influencing evolution originated in the early 20th century when the field of radiation biology was in its infancy, and our knowledge of the sources and effects of radiation was limited. Cosmic rays, which consist of high-energy particles, including protons and atomic nuclei, originate from various astrophysical sources such as the Sun and distant supernovae. They constantly bombard the Earth's atmosphere and surface.

The idea that cosmic rays could be involved in genetic mutations and evolution emerged as a speculative hypothesis rather than a well-developed theory with a single originator. Instead, it was the result of various scientists exploring the potential consequences of cosmic ray exposure on living organisms. The primary premise was that cosmic rays, with their high energy, could penetrate biological tissues and cause genetic changes in the form of mutations.

One of the earliest proponents of this hypothesis was Hermann Joseph Muller, an American geneticist who received the Nobel Prize in Physiology or Medicine in 1946 for his work on the induction of mutations by X-rays. Muller's research demonstrated that exposure to X-rays, a form of ionizing radiation, could induce mutations in the genes of fruit flies. Building on this work, he extended the idea to cosmic rays, suggesting that these high-energy particles from space could have similar mutagenic effects.

Muller's hypothesis about cosmic rays and mutation garnered attention and intrigue within the scientific community. At the time, the mechanisms of genetic mutation were not fully understood, and the idea that external factors, such as radiation, could induce genetic changes was a topic of active research and debate.

While Muller's work contributed to the hypothesis that cosmic rays might influence evolution, it is essential to note that his hypothesis was speculative, and he did not provide conclusive evidence to support the idea. Furthermore, the scientific community remained cautious and sought more definitive evidence before accepting cosmic rays as a significant driver of evolution.

The theory that cosmic rays play a substantial role in evolution faced challenges and eventual refutation as our understanding of genetics, radiation biology, and evolutionary processes advanced. As research in these fields progressed, it became clear that while cosmic rays indeed have the potential to cause genetic mutations, their contribution to the overall rate of mutations and evolution on Earth was relatively minor compared to other factors.

One crucial development that contributed to the refutation of the theory was the discovery of other, more potent sources of mutations. It became evident that the majority of mutations are spontaneous, arising from errors in DNA replication and repair processes within an organism's cells. These spontaneous mutations far outweighed the contribution of external factors like cosmic rays.

Additionally, scientists gained a better understanding of the mechanisms by which radiation-induced mutations occur. It became clear that the energy levels of cosmic rays were often insufficient to penetrate the protective layers of the Earth's atmosphere and reach the surface in significant quantities. Most cosmic rays are stopped by the atmosphere, reducing their impact on living organisms.

Furthermore, the theory faced challenges from studies that examined the actual rates of mutations and the influence of various radiation sources, including cosmic rays. These studies demonstrated that the overwhelming majority of mutations in the genetic material of organisms were due to intrinsic factors, and the contribution of cosmic rays was minimal.

As a result of these advancements in genetics and radiation biology, the theory of cosmic rays significantly influencing evolution lost support within the scientific community. While cosmic rays can induce mutations, they were no longer considered a primary driver of evolutionary change. Instead, evolutionary biologists focused on factors such as natural selection, genetic drift, and other intrinsic genetic processes as the primary drivers of evolution.

In summary, the hypothesis that cosmic rays could influence evolution emerged in the early 20th century as a speculative idea within the context of limited understanding of radiation biology and genetics. Hermann Joseph Muller, a Nobel laureate in physiology or medicine, played a role in popularizing this hypothesis, although it remained speculative in nature. As scientific research advanced, it became clear that cosmic rays, while capable of inducing mutations, had a relatively minor impact on the overall rate of mutations and evolution compared to other factors. Consequently, the theory of cosmic rays significantly influencing evolution lost support and relevance as our understanding of genetics and evolutionary processes evolved.

# Time Travel through Wormholes

The concept of time travel through wormholes, a fascinating and often explored idea in science fiction and theoretical physics, has captured the imagination of many. While it is an intriguing concept, it is currently considered highly unlikely and remains within the realm of theoretical speculation rather than established science. To understand why it is deemed unlikely, let's delve into the theory, its origins, and the scientific reasoning behind its skepticism.

The theory of time travel through wormholes involves the idea that traversable wormholes, hypothetical shortcuts or tunnels through spacetime, could be used as portals for time travel. According to this concept, if one could navigate through a wormhole, they might find themselves in a different region of spacetime, potentially at a different point in time.

The concept of wormholes itself is rooted in the mathematics of general relativity, Albert Einstein's theory of gravity, which describes the curvature of spacetime by massive objects. In essence, a wormhole is a theoretical tunnel connecting two separate points in spacetime, potentially allowing for shortcuts across vast cosmic distances.

The origins of the theory of wormholes can be traced back to the early 20th century when the mathematics of general relativity began to be explored more deeply. However, it was the physicist John Archibald Wheeler who popularized the term "wormhole" in the mid-20th century. Wheeler and his colleagues discussed these hypothetical structures as solutions to Einstein's field equations.

While the concept of wormholes intrigued scientists and writers alike, it was the renowned physicist Kip Thorne who contributed significantly to the theoretical groundwork and popularization of wormholes as potential means of time travel. Thorne, a Nobel laureate and a leading figure in astrophysics, explored the mathematics and physics of wormholes in his work. He suggested that if a traversable wormhole existed, it could theoretically allow for time travel under certain conditions.

Thorne's ideas, while influential, were theoretical in nature and did not provide a definitive path to time travel through wormholes. The scientific community recognized the immense theoretical challenges and paradoxes associated with this concept, leading to skepticism and caution.

One of the central reasons why time travel through wormholes is considered highly unlikely is the need for exotic matter. In order to stabilize a traversable wormhole and prevent its collapse, a form of exotic matter with negative energy density and tension would be required. This exotic matter would need to counteract the gravitational forces trying to crush the wormhole.

The concept of exotic matter with negative energy density raises significant theoretical and practical challenges. While such matter has been theorized in the context of exotic concepts like the Casimir effect and vacuum fluctuations, it remains speculative and has not been observed or harnessed in any practical sense. The existence of negative-energy exotic matter is not supported by our current understanding of physics.

Furthermore, the concept of time travel through wormholes introduces numerous paradoxes and logical inconsistencies. The most famous of these is the "grandfather paradox," which arises when a time traveler could potentially go back in time and prevent their own grandparents from meeting, leading to a self-contradictory situation.

Theoretical physics has explored these paradoxes in depth, leading to various proposed resolutions, such as the Novikov self-consistency principle, which suggests that any action taken in the past would already be accounted for in the timeline, preventing paradoxes. However, these resolutions are highly speculative and have not been tested or demonstrated in practice.

Another challenge to the feasibility of time travel through wormholes is the extreme conditions required to create and maintain these structures. Wormholes, according to current theoretical models, would likely be incredibly small and short-lived, making them impractical for human travel. Additionally, the energy requirements for manipulating and stabilizing wormholes are currently beyond our technological capabilities.

While the idea of time travel through wormholes remains a tantalizing and captivating concept in both science fiction and theoretical physics, it is important to emphasize that it is highly unlikely and speculative. Theoretical models involving wormholes require exotic matter with negative energy density, introduce paradoxes and logical inconsistencies, and rely on conditions that are currently beyond our technological reach.

In recent years, the focus of theoretical physics has shifted toward other avenues for understanding the nature of time, such as the study of black holes, the theory of relativity, and the exploration of the fundamental structure of the universe. While the concept of time travel through wormholes continues to capture the imagination, it remains a topic of scientific curiosity rather than a practical possibility.

In summary, the theory of time travel through wormholes is a captivating idea rooted in the mathematics of general relativity and explored by physicists like Kip Thorne. However, it is considered highly unlikely due to the theoretical challenges associated with stabilizing wormholes using exotic matter with negative energy density, the introduction of paradoxes, and the extreme conditions required for their creation and maintenance. While it remains a popular theme in science fiction, time travel through wormholes remains speculative and has not been demonstrated or realized in practice.

## Geology

## *The Lemurian Hypothesis: An Exploration of an Extinct Continent*

The Lemurian Hypothesis, a speculative theory, posits the existence of an ancient continent called Lemuria or Mu. This theory, originating in the late 19th century, suggests that Lemuria was a lost landmass located in the Indian and Pacific Oceans. Throughout its history, the Lemurian Hypothesis has captured the imaginations of both scientists and mystics, but it ultimately got disproved due to advances in geology and plate tectonics.

### The Essence of the Lemurian Hypothesis

The Lemurian Hypothesis proposes the existence of a continent known as Lemuria or Mu, which supposedly existed long before the continents we know today. This theoretical landmass was believed to have bridged the gap between the Indian and Pacific Oceans. The Lemurian Hypothesis suggested that Lemuria was home to an advanced and ancient civilization.

### Origins of the Lemurian Hypothesis

The Lemurian Hypothesis emerged in the late 19th century as part of the broader field of speculative theories about lost continents. It was developed as an attempt to explain the distribution of lemurs, a type of primate found in Madagascar and parts of Africa and India.

### Augustus Le Plongeon: A Key Proponent

Augustus Le Plongeon, a 19th-century amateur archaeologist and photographer, is often credited with popularizing the Lemurian Hypothesis. Le Plongeon's research in the Yucatan, particularly around the ancient Mayan city of Chichen Itza, led him to propose the idea of a lost continent that connected the Americas to India and the Pacific Islands.

**Believers in the Lemurian Hypothesis**

The Lemurian Hypothesis attracted a range of proponents, including both scientists and mystics, who were drawn to the idea of a lost, advanced civilization. Some of the notable figures who believed in the theory include:

**James Churchward:** An English author, James Churchward, played a significant role in popularizing the Lemurian Hypothesis. He claimed to have discovered ancient tablets in India that contained information about the lost continent of Lemuria.

**Madame Blavatsky:** Helena Petrovna Blavatsky, a prominent figure in the Theosophical Society, incorporated elements of the Lemurian Hypothesis into her teachings. Theosophy, a spiritual movement, embraced the idea of ancient, advanced civilizations.

**The Disproof of the Lemurian Hypothesis**

The Lemurian Hypothesis gradually lost credibility as scientific knowledge advanced. Several key factors contributed to its disproof:

**Advances in Geology:** The field of geology made significant progress in the understanding of continental drift and plate tectonics during the 20th century. These discoveries provided a more accurate explanation for the distribution of continents and ocean basins.

**Distribution of Lemurs:** The Lemurian Hypothesis was initially proposed to explain the distribution of lemurs. However, further research into the evolution and migration patterns of lemurs provided alternative, scientifically sound explanations for their presence in Madagascar and other regions.

**Lack of Geological Evidence:** Despite extensive exploration of the Indian and Pacific Oceans, no substantial geological evidence supporting the existence of a lost continent like Lemuria was found.

### The Timeline of Disproof

The disproof of the Lemurian Hypothesis occurred gradually throughout the 20th century:

**Early 20th Century:** The development of plate tectonics theory in the mid-20th century provided a comprehensive explanation for the movement of continents, rendering the concept of Lemuria unnecessary.

**Mid-20th Century:** Geological surveys of the Indian and Pacific Oceans failed to discover any geological remnants of a lost continent.

**Late 20th Century:** By the latter half of the 20th century, the Lemurian Hypothesis had lost its scientific credibility and was largely relegated to the realm of pseudoscience.

### Conclusion

The Lemurian Hypothesis, proposing the existence of a lost continent called Lemuria or Mu, captured the imaginations of scientists and mystics alike in the late 19th and early 20th centuries. Figures like Augustus Le Plongeon and James Churchward contributed to its popularity. However, as the field of geology advanced and plate tectonics theory emerged, the Lemurian Hypothesis lost credibility. Today, it serves as a historical example of a once-influential theory that was ultimately disproved by scientific progress and a deeper understanding of Earth's geological processes.

# The Expanding Earth Hypothesis

The Expanding Earth Hypothesis is a geological theory that suggests Earth was once significantly smaller and has gradually increased in size over geological time scales. This concept, while unconventional, has a unique history and attracted the attention of notable scientists. In this exploration, we will uncover the origins of the theory, its proponents, the challenges it faced, and when it was ultimately disproven.

**Unveiling the Expanding Earth Hypothesis**

The Expanding Earth Hypothesis rests on several key principles:

**Earth's Size Increase:** This theory posits that Earth has experienced substantial growth in its diameter over billions of years. This expansion is believed to have occurred mainly through the addition of new material at the mid-ocean ridges.

**Continental Drift:** The hypothesis is closely related to the concept of continental drift, which suggests that continents were once part of a supercontinent called Pangaea and have since moved apart.

**Origins of the Expanding Earth Hypothesis**

The Expanding Earth Hypothesis has an unconventional history:

**Samuel Warren Carey:** Australian geologist Samuel Warren Carey is often credited with formulating and popularizing the Expanding Earth Hypothesis. In the mid-20th century, Carey proposed that Earth's continents had drifted apart, and to accommodate this, Earth's size must have increased. He argued that new material was continuously added at the mid-ocean ridges, causing the planet to expand.

## Supporters and Believers

While the Expanding Earth Hypothesis did not gain widespread acceptance, it did attract the attention of some prominent scientists:

**Hans Pettersson:** Swedish geologist Hans Pettersson was one of the early supporters of the expanding Earth concept. He believed that the growth of Earth could explain various geological phenomena, including the formation of mountain ranges.

**Ott Christoph Hilgenberg:** German geophysicist Ott Christoph Hilgenberg was another proponent of the Expanding Earth Hypothesis. He developed a model of Earth expansion and argued that the increasing radius of the planet could account for continental drift.

## The Demise of the Expanding Earth Hypothesis

The Expanding Earth Hypothesis faced several significant challenges and ultimately lost favor within the scientific community:

**Lack of Mechanism:** One of the fundamental issues with the theory was the lack of a plausible mechanism to explain how Earth could expand. The idea that new material was continuously added at mid-ocean ridges was not supported by geological evidence.

**Plate Tectonics:** The emergence of plate tectonics in the mid-20th century provided a comprehensive and widely accepted explanation for continental drift and geological phenomena. Plate tectonics theory did not require Earth to expand, and it became the dominant framework in geology.

**Geological Evidence:** Geological evidence, including the distribution of fossils and rock formations, did not align with the predictions of the Expanding Earth Hypothesis. Instead, plate tectonics provided a more consistent and evidence-based explanation for Earth's geological history.

**Global Positioning System (GPS):** The development and widespread use of GPS technology allowed scientists to precisely measure the motion of tectonic plates and continental drift. These measurements confirmed the predictions of plate tectonics and further discredited the Expanding Earth Hypothesis.

### When the Expanding Earth Hypothesis Got Disproved

The Expanding Earth Hypothesis gradually lost credibility throughout the latter half of the 20th century and can be considered largely disproved by that time. The widespread acceptance of plate tectonics and the accumulation of evidence supporting this theory led to the rejection of the expanding Earth concept within the scientific community.

In conclusion, the Expanding Earth Hypothesis proposed that Earth's diameter had increased significantly over geological time scales, primarily through the addition of new material at mid-ocean ridges. Samuel Warren Carey and other scientists played a role in formulating and advocating for this theory. However, the hypothesis faced substantial challenges, including a lack of a plausible mechanism, the emergence of plate tectonics, and geological evidence that contradicted its predictions. As plate tectonics became the dominant framework in geology and technology like GPS provided supporting data, the Expanding Earth Hypothesis gradually lost favor and can be considered largely disproved by the latter half of the 20th century.

# Geosynclinal Theory

The geosynclinal theory was a significant concept in the history of geology that sought to explain the formation of mountain ranges and the Earth's crustal movements. It originated in the 19th century and underwent various modifications over time. This theory, though influential for many years, eventually faced scientific scrutiny and was replaced by the theory of plate tectonics.

**Understanding the Geosynclinal Theory**

The geosynclinal theory can be broken down into several key components:

**Formation of Sediments:** The theory postulated that sediments were continually deposited in vast oceanic depressions, known as geosynclines, over millions of years.

**Compression and Uplift:** As sediments accumulated in these geosynclines, they experienced tremendous pressure and heat. Over time, this pressure resulted in the folding and uplifting of the sediments, ultimately forming mountain ranges.

**Cyclical Process:** The geosynclinal theory suggested that this process was cyclical, with geosynclines forming, filling with sediments, and then being compressed and uplifted to create mountain ranges.

## Origins of the Geosynclinal Theory

The geosynclinal theory has its origins in the 19th century:

**James Hall:** James Hall, an American geologist, is often credited with the initial development of the geosynclinal concept in the mid-19th century. He proposed that large troughs in the Earth's crust, the geosynclines, were responsible for the formation of mountain ranges and the deposition of sediments.

## Supporters and Believers

The geosynclinal theory gained considerable support among geologists and scientists during the late 19th and early 20th centuries:

**James Dana:** An influential American geologist, James Dana, expanded on Hall's ideas and further developed the geosynclinal theory in his work "Manual of Geology" (1863). Dana's efforts helped solidify the theory's acceptance within the geological community.

**John Wesley Powell:** Powell, another American geologist, conducted extensive surveys of the American West and proposed that the geosynclinal theory explained the complex geological features of the region.

**Alexander von Humboldt:** The renowned German naturalist and explorer, Humboldt, provided support for the theory by suggesting that geosynclines played a role in the formation of mountain systems.

## The Demise of the Geosynclinal Theory

The geosynclinal theory was widely accepted and influential for many years. However, it eventually faced several challenges and was replaced by the theory of plate tectonics:

**Plate Tectonics Emerges:** In the mid-20th century, the theory of plate tectonics revolutionized the field of geology.

It provided a more comprehensive and scientifically sound explanation for the movement of Earth's crustal plates, the formation of mountain ranges, and various geological phenomena.

**Evidence Accumulates:** Plate tectonics was supported by a wealth of empirical evidence, including the mapping of mid-ocean ridges, the discovery of subduction zones, and the ability to explain seismic activity and volcanism. This accumulating evidence undermined the geosynclinal theory.

**Plate Tectonics Unification:** Unlike the geosynclinal theory, which focused primarily on mountain building, plate tectonics provided a unified explanation for a wide range of geological processes, including the movement of continents, the creation of ocean basins, and the formation of mountain ranges.

### When the Geosynclinal Theory Got Disproved

The geosynclinal theory gradually lost favor among scientists during the mid-20th century:

**1960s and 1970s:** The emergence and widespread acceptance of the theory of plate tectonics in the 1960s and 1970s marked the decline of the geosynclinal theory. Plate tectonics provided a more comprehensive and scientifically robust framework for understanding Earth's geological processes.

In summary, the geosynclinal theory was a 19th-century concept that aimed to explain the formation of mountain ranges and the Earth's crustal movements. It proposed that vast oceanic depressions, called geosynclines, collected sediments over millions of years, which were then compressed and uplifted to create mountains. The theory was developed by geologists like James Hall and gained support from prominent figures such as James Dana and Alexander von Humboldt.

However, it eventually fell out of favor as the theory of plate tectonics emerged with a more comprehensive and evidence-based explanation for geological phenomena. The demise of the geosynclinal theory marked a significant milestone in the history of geology, highlighting the evolving nature of scientific understanding.

# The Cosmic Ice Theory

The cosmic ice theory, also known as the ice comet theory, was a unique concept proposed to explain the origins of comets and their icy compositions. This theory emerged during the late 19th century and underwent revisions and debates among scientists. Although it gained some initial support, it eventually gave way to a more comprehensive understanding of comets and their formation.

**Understanding the Cosmic Ice Theory**

The cosmic ice theory can be summarized as follows:

**Comet Origins:** The theory posited that comets originated from vast reservoirs of cosmic ice located at the outer reaches of the solar system.

**Solar Heating:** As comets approached the Sun in their elliptical orbits, the intense solar radiation caused the cosmic ice to vaporize and release gas and dust, forming the characteristic tails of comets.

**Replenishment:** According to the cosmic ice theory, comets were not a finite resource but rather a continuous process. As comets lost material during their passages near the Sun, new comets were thought to form from the remaining cosmic ice, ensuring a constant supply of comets in the solar system.

**Origins of the Cosmic Ice Theory**

The cosmic ice theory can be traced back to the late 19th century:

**Friedrich Wilhelm Bessel:** Although the idea of comets originating from cosmic ice was not initially formulated by a single individual, Friedrich Wilhelm Bessel, a German mathematician and astronomer, contributed to the early discussions surrounding comet origins in the early 19th century.

**Ernst Friedrich Wilhelm Klinkerfues:** Klinkerfues, another German astronomer, proposed a variation of the cosmic ice theory in 1872, suggesting that comets were composed of solid chunks of ice.

**Supporters and Believers**

The cosmic ice theory garnered some support and interest among scientists during the late 19th and early 20th centuries:

**Giovanni Schiaparelli:** The Italian astronomer Giovanni Schiaparelli conducted studies on comets and expressed support for the cosmic ice theory.

**William Henry Pickering:** The American astronomer William Henry Pickering also explored the idea of cosmic ice as a source for comets.

**The Demise of the Cosmic Ice Theory**

While the cosmic ice theory captured the imagination of some astronomers, it ultimately faced challenges and was supplanted by a more comprehensive understanding of comets:

**Spectroscopic Analysis:** Advancements in spectroscopy allowed scientists to analyze the composition of comets more accurately. These studies revealed that comets contained a wide range of volatile substances, including water, carbon dioxide, and various organic compounds. These findings contradicted the notion of comets being primarily composed of solid ice.

**Comet Missions:** Space missions, such as NASA's Stardust and ESA's Rosetta, provided direct observations of comets and their nuclei. These missions confirmed that comets have complex compositions that include both ice and dust. The notion of solid chunks of ice making up comets was largely debunked.

## When the Cosmic Ice Theory Got Disproved

The cosmic ice theory gradually fell out of favor during the early 20th century:

**1900s and 1910s:** As spectroscopic analysis and the study of comets' compositions advanced, it became evident that comets were not primarily composed of solid ice. This realization led to the decline of the cosmic ice theory.

In summary, the cosmic ice theory, which proposed that comets originated from vast reservoirs of cosmic ice and released gas and dust as they approached the Sun, emerged during the late 19th century. Notable figures like Friedrich Wilhelm Bessel and Ernst Friedrich Wilhelm Klinkerfues contributed to its development. Some astronomers, including Giovanni Schiaparelli and William Henry Pickering, expressed support for this theory. However, the theory faced challenges as spectroscopic analysis and space missions provided more accurate insights into comet compositions. These advancements ultimately led to the rejection of the cosmic ice theory, making way for a more comprehensive understanding of comets and their complex compositions.

# Catastrophism

Catastrophism was a prominent geological theory that postulated that the Earth's geological features and the history of life were primarily shaped by sudden and catastrophic events. This theory, which emerged in the late 18th century, stood in stark contrast to the prevailing idea of uniformitarianism, which posited that geological processes occurred gradually and uniformly over vast periods of time. Catastrophism had a significant impact on early geological thought and was supported by some notable scientists, including Georges Cuvier. However, as the science of geology advanced and a deeper understanding of Earth's processes emerged, catastrophism gradually lost favor. In this exploration, we will delve into the essence of catastrophism, its historical origins, its early proponents, the influential individuals who believed in its validity, the process by which it was ultimately disproved, and the timeline of its discrediting.

## The Essence of Catastrophism

Catastrophism proposed that Earth's geological features, such as mountains, valleys, and the fossil record, were primarily the result of sudden and dramatic events, such as floods, earthquakes, and volcanic eruptions. These catastrophic events were believed to have occurred intermittently and had a profound impact on the planet's surface and the history of life.

### Origins of Catastrophism

The origins of catastrophism can be traced to the late 18th century, a period marked by significant developments in geological thinking.

**Georges Cuvier:** Georges Cuvier, a pioneering French naturalist and paleontologist, is often credited with formulating and popularizing the concept of catastrophism. In the early 19th century, Cuvier examined fossils and proposed that the Earth had experienced a series of catastrophic events that led to the extinction of certain species.

### Proponents of Catastrophism

Catastrophism gained support from notable scientists of the time, including Cuvier and other early geologists.

**Cuvier's Extinction Hypothesis:** Cuvier's work on the extinction of species, based on his examination of fossils, was a cornerstone of catastrophism. He argued that a series of catastrophic events had led to the abrupt disappearance of certain species, followed by the creation of new species.

**James Hutton's Uniformitarianism:** While James Hutton is often associated with uniformitarianism, his early work also contained elements of catastrophism. He recognized that catastrophic events like floods and volcanic eruptions had played a role in shaping the Earth's surface.

### The Decline of Catastrophism: Emergence of Modern Geology

The credibility of catastrophism began to wane as the science of geology matured and a more comprehensive understanding of Earth's processes emerged. Several key developments contributed to the discrediting of the theory.

**Uniformitarianism Gains Prominence:** The idea of uniformitarianism, proposed by James Hutton and later popularized by Charles Lyell in the early 19th century, began to gain prominence. Uniformitarianism posited that geological processes occurred gradually and uniformly over long periods of time, challenging the catastrophic explanations proposed by catastrophism.

**Charles Lyell's Principles of Geology:** Charles Lyell's influential work, "Principles of Geology," published in the 1830s, provided a comprehensive argument for uniformitarianism. Lyell argued that the same geological processes observed today had been shaping the Earth's surface throughout its history, gradually and continuously.

**Gradualism and Evolutionary Theory:** Charles Darwin's theory of evolution, published in the mid-19th century, further bolstered the idea of gradual change in the natural world. It provided an alternative explanation for the development of life that did not require catastrophic events.

**When Catastrophism Was Disproved**

The decline and discrediting of catastrophism occurred gradually over the course of the 19th century, marking a significant shift in geological thinking.

**Early 19th Century:** Georges Cuvier's work on extinction was influential but became increasingly marginalized as uniformitarianism gained ground.

**Mid-19th Century:** Charles Lyell's "Principles of Geology" provided a compelling argument for uniformitarianism, and his ideas became widely accepted within the geological community.

**Late 19th Century:** The advent of evolutionary theory and the development of modern geology, which emphasized gradual processes, further marginalized catastrophism.

**Legacy and Insights**

Catastrophism left an indelible mark on the history of geology, serving as a transitional stage in the development of geological thought. While it may have been superseded by uniformitarianism and modern geology, it contributed to the broader understanding of Earth's history and the recognition that both gradual and catastrophic processes have played roles in shaping the planet. Catastrophism's legacy lies in its role as a stepping stone in the evolution of scientific thought, illustrating how scientific paradigms can shift and evolve over time as new evidence and ideas emerge.

# VI

# Psychology

# The "Blank Slate" Theory

The "Blank Slate" theory, also known as the "tabula rasa" theory, was a concept deeply rooted in the history of philosophy and psychology. It posited that individuals are born with a blank mental state, devoid of innate traits, characteristics, or predispositions, and that all knowledge and behaviors are acquired through experience and environmental influences. This theory had a profound impact on the way scholars and thinkers approached the understanding of human nature for centuries. In this exploration, we will delve into the origins of the "Blank Slate" theory, its proponents, the era in which it gained prominence, the eventual debunking of the theory, and the impact of its discrediting on our understanding of human nature.

**Origins of the Theory:** The "Blank Slate" theory, or tabula rasa in Latin, can be traced back to ancient Greek philosophers, particularly the works of Aristotle and Plato. Aristotle, for instance, posited that the mind of a child at birth was like a blank tablet, ready to be filled with knowledge and experiences. However, the theory gained prominence and became a central concept in modern philosophy and psychology during the Enlightenment era.

**John Locke and the Enlightenment Era:** The "Blank

Slate" theory, as it is commonly understood today, owes much to the Enlightenment philosopher John Locke. In his influential work, "An Essay Concerning Human Understanding" (1690), Locke argued that the human mind is a blank slate at birth, devoid of innate ideas, and that knowledge and character are formed through sensory experiences and social interactions.

Locke's ideas had a profound impact on Enlightenment thinking, particularly on the development of empiricism—the belief that knowledge is derived from sensory experiences. This theory also laid the foundation for the "nurture over nature" perspective, suggesting that environmental factors shape human behavior and identity.

**Prominent Believers and Influence:** The "Blank Slate" theory gained widespread acceptance and influenced prominent thinkers in various fields, not limited to philosophy and psychology. It became a foundational concept in education, social sciences, and political philosophy. Some of the notable figures who endorsed or were influenced by this theory include:

**Jean-Jacques Rousseau:** The Enlightenment philosopher Rousseau built upon the "Blank Slate" theory in his work, particularly in his treatise "Emile, or On Education" (1762). He argued that children are born inherently good and that society corrupts their natural goodness.

**Behaviorist Psychologists:** In the early 20th century, behaviorist psychologists such as John B. Watson and B.F. Skinner were proponents of the idea that human behavior is largely a product of environmental conditioning, further emphasizing the "Blank Slate" perspective.

**Challenges to the Theory:** The "Blank Slate" theory

faced increasing challenges as scientific and philosophical understanding evolved. Several key factors contributed to the eventual debunking of the theory:

**Advancements in Genetics:** The discovery of genetics and the understanding of hereditary traits challenged the notion of a completely blank mental state at birth. It became evident that some aspects of human behavior and characteristics are influenced by genetic factors.

**Cognitive Psychology:** The emergence of cognitive psychology in the mid-20th century shifted the focus from behaviorist perspectives to the study of mental processes. This field revealed that cognitive development and certain cognitive abilities have innate components.

**Ethology and Evolutionary Psychology:** Ethology, the study of animal behavior, and evolutionary psychology emphasized the role of evolution and innate predispositions in shaping human behavior and cognition. The study of instincts and evolutionary adaptations challenged the "Blank Slate" perspective.

**The Debunking of the Theory:** The "Blank Slate" theory was gradually debunked as empirical evidence accumulated in various fields. The concept of innate traits, including cognitive capacities, emotions, and certain predispositions, gained acceptance within psychology and the natural sciences.

By the late 20th century, the theory was no longer considered a tenable explanation for human nature. The fields of genetics, neuroscience, cognitive psychology, and evolutionary psychology had provided substantial evidence that humans are not born as blank slates but have innate characteristics and predispositions that influence their development and behavior.

**Modern Understanding of Human Nature:** The debunking of the "Blank Slate" theory paved the way for a more nuanced and comprehensive understanding of human nature. Contemporary research recognizes that both nature (biological and genetic factors) and nurture (environmental influences) interact in complex ways to shape human development, behavior, and identity.

This modern perspective acknowledges that humans are not completely malleable at birth, nor are they predetermined by their genetics. Instead, they inherit a range of predispositions and potentials, which interact with their environment, experiences, and culture to form their unique individualities.

**Conclusion:** The "Blank Slate" theory, once a central concept in philosophy, psychology, and education, has been debunked through centuries of evolving scientific and philosophical thought. It was a product of its time, a reflection of the Enlightenment era's emphasis on reason, empiricism, and the power of experience. However, as knowledge in genetics, neuroscience, cognitive psychology, and evolutionary psychology advanced, the theory gradually lost its credibility.

The discrediting of the "Blank Slate" theory marked a pivotal moment in the history of understanding human nature, emphasizing the complex interplay of innate traits and environmental influences in shaping human behavior and identity. It also paved the way for more nuanced and evidence-based approaches in psychology, philosophy, and the social sciences.

# Phrenology

Phrenology was a pseudoscientific theory that claimed to be able to discern an individual's personality traits, intellect, and moral character by examining the shape and bumps of their skull. This now-discredited theory emerged in the late 18th century and gained popularity in the 19th century. It was based on the idea that different regions of the brain controlled specific mental faculties and that the size and shape of these regions could be inferred from the external features of the skull. Although phrenology was widely accepted by some during its heyday, it eventually fell into disrepute due to scientific scrutiny and lack of empirical evidence. In this exploration, we will delve into what phrenology entailed, when it first emerged, its origins, the key proponents who championed it, the gradual erosion of its credibility, and the pivotal moment when phrenology was conclusively disproven.

## The Essence of Phrenology

Phrenology was built upon the idea that the brain was divided into distinct areas, each responsible for a particular mental function or personality trait. These regions were believed to exert pressure on the inside of the skull, creating bumps and indentations that could be felt and analyzed. Phrenologists claimed that by measuring these cranial features, they could reveal an individual's character and abilities.

## Origins and Early Notions of Phrenology

The origins of phrenology can be traced to Franz Joseph Gall, an Austrian physician and anatomist who lived in the late 18th and early 19th centuries. Gall developed a theory that different mental faculties were localized in specific areas of the brain. He believed that the development of these brain areas influenced the shape of the skull and that one could infer a person's mental traits by examining these cranial features.

## Development of Phrenological Doctrine

Gall's ideas laid the groundwork for the development of phrenology as a formal doctrine. It was Gall's student, Johann Spurzheim, who popularized the practice and introduced the term "phrenology" to describe the study of character through cranial examination. Phrenology gained prominence in the early 19th century, spreading throughout Europe and the United States.

## Proponents of Phrenology

Phrenology attracted the interest of a range of individuals, including scientists, medical practitioners, and laypeople. Its proponents believed that it offered a systematic way to understand human behavior and personality.

**George Combe's Influence:** George Combe, a Scottish lawyer, was one of the most influential proponents of phrenology. His book "The Constitution of Man" published in 1828, contributed to the popularization of phrenology and its spread to the United States.

## The Decline of Phrenology: Empirical Challenges

Phrenology began to face increasing skepticism and scientific scrutiny as the 19th century progressed. Critics argued that the theory lacked empirical evidence and that its claims were based on anecdotal observations rather than controlled experiments.

**Scientific Rejection:** Phrenology's claims were met with skepticism from the scientific community. Prominent scientists and anatomists, including Pierre Flourens and Marie-Jean-Pierre Flourens, conducted experiments and dissections that failed to support phrenology's claims about localized brain functions.

**Challenges to Phrenological Practice:** Phrenologists often engaged in questionable practices, such as performing skull readings on criminals and the mentally ill. These practices were criticized for their lack of scientific rigor and ethics.

### The Advent of Modern Neuroscience

The ultimate downfall of phrenology was precipitated by the development of modern neuroscience in the late 19th and early 20th centuries. Advances in neuroanatomy and physiology revealed a vastly more complex and nuanced understanding of brain function.

**Localization of Function:** Research in neuroscience demonstrated that mental functions were not localized in the simplistic manner proposed by phrenology. Instead, mental processes were found to involve intricate networks of brain regions.

**Broca's Discovery:** Paul Broca's discovery in the mid-19th century of the brain region responsible for language (now known as Broca's area) was a significant blow to phrenology. This discovery showed that specific brain functions were not tied to the external features of the skull but could be identified through careful anatomical studies.

## Phrenology's Discrediting and Legacy

By the late 19th century, phrenology had largely fallen into disrepute within the scientific community. It was dismissed as a pseudoscience, and its claims were debunked. Phrenology's decline marked a transition from an era of speculative theories about the brain to the development of modern neuroscience based on empirical evidence and rigorous experimentation.

In summary, phrenology, the pseudoscientific theory of reading personality traits and mental abilities through cranial features, emerged in the late 18th century and gained popularity in the 19th century. It was based on the erroneous belief that the brain's functions were localized in distinct regions that could be inferred from skull shape. Despite having influential proponents, phrenology was gradually discredited as advances in neuroscience revealed the true complexity of brain function. By the late 19th century, it had lost scientific credibility, and it serves today as a historical example of a once-prominent but ultimately flawed pseudoscience.

# Hysteria as a Wandering Uterus

Hysteria as a wandering uterus is a theory deeply rooted in the history of medicine and psychology, representing a captivating but profoundly misguided belief about the female body. This theory, which suggested that various emotional and psychological disturbances in women were attributed to the movement of the uterus within the body, has a rich history that spans centuries. In this exploration, we will delve into the origins of this theory, its proponents, the enduring belief in its validity, and the eventual debunking of this notion.

**Origins of the Theory:** The concept of hysteria as a wandering uterus can be traced back to ancient Greece, where it was foundational in the understanding of women's health. The term "hysteria" itself derives from the Greek word "hystera," meaning uterus. The theory posited that the uterus, considered a mobile organ, could move within a woman's body, causing a range of psychological and physical symptoms.

**Ancient Beliefs:** Hippocrates, often referred to as the father of medicine, is one of the earliest known proponents of the theory. In his writings, he attributed various ailments, particularly those related to emotional and mental distress, to the wandering uterus. Hippocrates believed that the uterus would travel throughout the female body, causing symptoms such as anxiety, irritability, and even suffocation.

The notion of a wandering uterus continued to gain prominence in ancient Greece and Rome, and it was widely accepted as the prevailing explanation for women's emotional and psychological issues during that era.

**Enduring Belief:** The theory of hysteria as a wandering uterus endured for centuries, shaping medical and societal perceptions of women's health. It became deeply ingrained in medical literature and was commonly accepted as an explanation for various female maladies, including mood swings, anxiety, and unexplained physical symptoms.

Throughout the Middle Ages and the Renaissance, this theory remained influential. Medical treatments aimed at addressing the wandering uterus included fumigations, dietary interventions, and even the application of pungent substances to encourage the uterus to return to its proper position. The belief in the wandering uterus persisted into the early modern period, reinforcing the idea that women's emotional well-being was closely tied to the position of their reproductive organs.

**Debunking the Theory:** The eventual debunking of the theory of hysteria as a wandering uterus was a gradual process that unfolded over centuries. Several key factors contributed to its eventual discrediting:

**Advancements in Anatomy:** As the field of anatomy advanced, a more accurate understanding of the female reproductive system emerged. It became evident that the uterus was not a mobile organ that could wander within the body, but rather a fixed and anchored structure.

**Medical Enlightenment:** The Enlightenment era ushered

in a more critical and empirical approach to medicine and science. Physicians and scholars began to question long-held beliefs and sought evidence-based explanations for medical phenomena.

**Emergence of Psychology:** The rise of psychology as a distinct field of study in the 19th century led to a shift in the understanding of mental health. Emotional and psychological disturbances were increasingly viewed through the lens of psychology and neurology rather than attributed solely to reproductive organs.

**Feminist Movement:** The feminist movement of the 20th century challenged prevailing notions about women's health and emphasized the need for gender-neutral and evidence-based medical practices. This advocacy contributed to a reevaluation of past beliefs and practices, including the theory of hysteria.

**The Discrediting of the Theory:** The theory of hysteria as a wandering uterus gradually lost credibility over several centuries, but it was not a sudden or dramatic discrediting. Instead, it was a gradual process influenced by changing paradigms in medicine, science, and societal perspectives on women's health.

By the mid-20th century, the theory had largely been abandoned by the medical community. Advancements in anatomy, the development of psychology as a discipline, and a more comprehensive understanding of mental health had rendered the theory obsolete. Medical practitioners and scholars came to recognize that women's emotional and psychological experiences were not solely determined by the position of their uterus but were influenced by a complex interplay of physiological, psychological, and social factors.

**In Conclusion:** The theory of hysteria as a wandering uterus, which persisted for centuries, represents a remarkable example of how deeply ingrained and misguided beliefs about women's health can be. It originated in ancient Greece and found support among influential figures in the history of medicine, including Hippocrates. The theory endured for centuries, shaping medical practices and societal perceptions of women's emotional well-being.

However, the gradual discrediting of this theory was driven by advancements in anatomy, the emergence of psychology, the Enlightenment era's empirical approach to medicine, and the advocacy of the feminist movement. By the mid-20th century, the theory had lost its validity and was consigned to the annals of medical history, highlighting the importance of evidence-based approaches in understanding and addressing women's health.

# The Refrigerator Mother Theory

The Refrigerator Mother Theory, a now-discredited explanation for the development of autism, stands as a sobering reminder of how misguided theories can have lasting consequences in the field of psychology and medicine. This theory, which posited that autism was caused by emotionally cold and distant mothers, had profound implications for both the understanding of the condition and the lives of affected families. In this exploration, we will delve into the origins of the Refrigerator Mother Theory, its proponents, the harm it inflicted, and the gradual process of its disproof.

**Origins of the Refrigerator Mother Theory:** The Refrigerator Mother Theory emerged in the mid-20th century as an attempt to explain the behavior and development of children later diagnosed with autism spectrum disorders. At its core, the theory suggested that autism was a result of maternal emotional neglect, particularly the mother's supposed inability to form a loving and nurturing bond with her child.

**Leo Kanner and the Early Years:** Dr. Leo Kanner, a prominent child psychiatrist, is often associated with the initial formulation of the Refrigerator Mother Theory. In 1943, he published a groundbreaking paper introducing the concept of "early infantile autism." In his paper, Kanner described a group of children who exhibited social withdrawal, communication difficulties, and repetitive behaviors. Although Kanner did not explicitly blame mothers, his descriptions of these children's parents, particularly the mothers, implied a lack of warmth and emotional connection.

**Prominent Believers and Widespread Acceptance:** The Refrigerator Mother Theory gained widespread acceptance within the medical and psychological communities, primarily due to Leo Kanner's influence. Many clinicians and researchers adopted this theory as a plausible explanation for autism. Some believed that cold and emotionally distant mothers were the cause of their children's developmental challenges. This belief not only stigmatized mothers but also diverted attention away from exploring the true underlying causes of autism.

**The Impact on Families:** The acceptance of the Refrigerator Mother Theory had profound and devastating consequences for families affected by autism. Parents, especially mothers, were unfairly blamed for their children's condition. They were often made to feel guilty and responsible for their child's struggles, adding an unnecessary burden to an already challenging situation. Families were isolated, and support for individuals with autism was limited because the prevailing theory suggested that emotional warmth and maternal bonding could "cure" the condition.

**The Slow Discrediting of the Theory:** The eventual discrediting of the Refrigerator Mother Theory was a slow and challenging process that unfolded over several decades. Several key factors contributed to its downfall:

**Emergence of Autism Advocacy:** Parents and advocacy

groups began to challenge the prevailing theory. They advocated for a more compassionate and understanding approach to autism and emphasized the need for research into the biological and genetic factors that contribute to the condition.

**Advancements in Autism Research:** Scientific research into autism began to uncover evidence of genetic and neurological factors associated with the condition. This shift in focus away from maternal blame helped erode the theory's credibility.

**Autism as a Spectrum:** The recognition that autism is a spectrum disorder with a wide range of presentations and causes made it increasingly clear that there could not be a single, maternal cause for all cases of autism.

**Diagnostic Changes:** Changes in diagnostic criteria and a broader understanding of autism led to a more inclusive view of the condition. It became evident that autism was not solely related to parenting practices.

**The Discrediting of the Theory:** By the late 20th century and into the 21st century, the Refrigerator Mother Theory had lost its credibility within the scientific and medical communities. Researchers, clinicians, and advocates had shifted their focus to understanding the complex genetic, neurological, and environmental factors that contribute to autism. The theory was no longer considered a valid explanation for the condition.

**Legacy and Lessons Learned:** The Refrigerator Mother Theory leaves a lasting legacy of harm and misunderstanding in the history of autism. It serves as a stark reminder of the consequences of blaming parents for their children's developmental challenges and the importance of evidence-based approaches in the field of psychology and medicine.

In conclusion, the Refrigerator Mother Theory, which erroneously posited that emotionally distant mothers were responsible for their children's autism, had its origins in the mid-20th century and was associated with Dr. Leo Kanner. The theory gained widespread acceptance, stigmatizing families and mothers affected by autism. It took several decades, the efforts of advocacy groups, advancements in research, and a shift in diagnostic and scientific understanding to discredit the theory. By the late 20th century, the theory had lost its credibility, emphasizing the need for compassion, support, and evidence-based approaches in the field of autism research and treatment.

# Lobotomy

Lobotomy, a discredited procedure in the history of psychology, represents a stark example of how medical and psychological theories can lead to misguided practices. This controversial procedure involved the surgical removal or disconnection of specific brain tissues and was once believed to be a viable treatment for various mental health conditions. This explanation will delve into the origins of the lobotomy theory, its proponents, the procedure's rise to popularity, and ultimately, its downfall.

The Origins of the Lobotomy Theory: The lobotomy theory originated in the early 20th century and was closely tied to the broader field of psychosurgery. It was developed as an attempt to find a surgical solution for individuals suffering from severe mental illnesses, particularly those deemed as incurable or highly disruptive to society. Although the precise origins are challenging to pinpoint, it gained prominence in the 1930s and 1940s.

The Pioneer: Dr. Egas Moniz: Dr. António Egas Moniz, a Portuguese neurologist and psychiatrist, is often credited as one of the key figures behind the development of the lobotomy procedure. In 1935, he introduced a technique called "leucotomy," which involved cutting white matter tracts in the frontal lobes of the brain to treat mental disorders. Moniz received the Nobel Prize in Physiology or Medicine in 1949 for his work on prefrontal leucotomy, despite the mounting controversy surrounding the procedure.

Prominent Supporters: The lobotomy procedure gained significant attention and support from some prominent scientists and medical professionals of the time. Walter J. Freeman, an American neurologist, and James W. Watts, a neurosurgeon, were among the early enthusiasts of the procedure. Freeman, in particular, played a crucial role in popularizing the lobotomy in the United States.

The Procedure's Rise to Popularity: Lobotomy's rise to popularity can be attributed to the belief that it offered a solution for individuals suffering from severe mental illnesses, such as schizophrenia and major depression, when other treatments had failed. The procedure was relatively simple and could be performed in outpatient settings, making it accessible to a wide range of patients.

The procedure involved drilling holes in the skull or using a thin instrument inserted through the eye sockets to sever or remove connections in the frontal lobes of the brain. The rationale was that disrupting these neural pathways would alleviate symptoms and reduce the severity of mental disorders. Lobotomies were performed on thousands of patients in the United States and Europe during the 1940s and 1950s.

The Downfall of the Lobotomy Theory: Despite its initial popularity, the lobotomy theory eventually faced significant opposition and scrutiny. Several factors contributed to its downfall:

Ethical Concerns: Lobotomy procedures often resulted

in severe cognitive and emotional impairments in patients. Critics argued that the procedure caused more harm than good and raised serious ethical questions about its use.

Lack of Scientific Evidence: The lobotomy procedure lacked empirical support and was based on anecdotal evidence rather than rigorous scientific research. As the field of psychology and neuroscience advanced, there was growing skepticism about the efficacy of lobotomies.

Emergence of Alternative Treatments: With the development of new psychotropic medications and the advent of psychotherapy, alternative treatments for mental disorders began to gain prominence. These treatments were seen as less invasive and more effective than lobotomies.

Changing Attitudes: As public awareness grew about the potential risks and adverse effects of lobotomies, attitudes toward the procedure shifted. The media played a role in highlighting the negative outcomes of lobotomies, contributing to a decline in public support.

The Disproval of the Lobotomy Theory: The discrediting of the lobotomy theory was a gradual process that unfolded over several decades. By the 1950s and 1960s, the procedure had already fallen out of favor, and its use had declined significantly. This decline was driven by a combination of factors, including increased ethical scrutiny, the emergence of alternative treatments, and a growing body of evidence that highlighted the procedure's shortcomings.

By the late 20th century, lobotomies had largely been abandoned as a psychiatric treatment. The development of more effective and less invasive therapies, coupled with a better understanding of the brain's complexity, contributed to the discrediting of the procedure. In modern times, lobotomies are viewed as a dark chapter in the history of psychology and medicine, serving as a cautionary tale about the dangers of adopting unproven and invasive treatments for mental disorders.

In conclusion, the lobotomy theory, which involved surgical interventions on the brain to treat mental illnesses, had its origins in the early 20th century and gained popularity in the mid-20th century. Dr. Egas Moniz, along with other prominent supporters, played a key role in its development and promotion. However, the procedure's downfall came about due to ethical concerns, lack of scientific evidence, the emergence of alternative treatments, and changing attitudes within the medical and psychiatric communities. While the lobotomy theory was once considered a valid approach, it has since been thoroughly discredited, serving as a stark reminder of the importance of evidence-based practices in the field of psychology and medicine.

# Primal Scream Therapy

Primal Scream Therapy, a psychological theory and therapeutic approach that gained notoriety in the 1970s, is a remarkable example of unconventional and unproven methods in the field of psychology. This therapy, developed by Arthur Janov, centered on the idea that emotional and psychological healing could be achieved by revisiting and expressing deep-seated childhood traumas through primal screams. In this exploration, we will delve into the origins of Primal Scream Therapy, the individuals who endorsed its effectiveness, the skepticism it faced, and its eventual decline as a widely accepted therapeutic approach.

**Origins of Primal Scream Therapy:** Primal Scream Therapy emerged in the 1960s and was pioneered by Arthur Janov, a clinical psychologist. The theory behind this approach was rooted in the belief that unresolved childhood traumas were the root cause of many mental health issues in adults. Janov proposed that by revisiting these traumatic experiences and expressing the suppressed emotions associated with them, individuals could achieve profound emotional and psychological healing.

**Arthur Janov:** Arthur Janov was the driving force behind Primal Scream Therapy. Born in 1924, he developed this therapy after years of working with patients in traditional psychotherapy settings. Janov believed that conventional talk therapy was insufficient in addressing the deep-seated emotional pain that he believed was at the core of many mental health problems. He advocated for a more direct and emotionally expressive approach to therapy, which eventually led to the development of Primal Scream Therapy.

**Prominent Supporters:** Primal Scream Therapy garnered attention and support from a range of individuals, including celebrities, writers, and psychologists. Notable figures such as John Lennon and Yoko Ono, members of The Beatles, were among the therapy's high-profile advocates. They famously underwent Primal Scream Therapy in the early 1970s and incorporated elements of it into their music and art.

The therapy also attracted interest from the counterculture movement of the 1960s and 1970s, which often embraced unconventional and alternative approaches to various aspects of life, including psychology and self-help.

**The Theory's Appeal:** The appeal of Primal Scream Therapy lay in its promise of catharsis and emotional release. Janov claimed that by allowing individuals to express their suppressed emotions, often through intense and guttural screams, they could break free from the emotional baggage of their past and achieve a state of emotional well-being. This approach seemed attractive to many who had struggled with traditional psychotherapy or were seeking rapid and dramatic transformations in their lives.

**Skepticism and Criticism:** Despite the enthusiasm and support it received, Primal Scream Therapy faced significant skepticism and criticism from within the psychological and medical communities. Some of the key criticisms included:

**Lack of Empirical Evidence:** Critics argued that Primal Scream Therapy lacked empirical evidence to support its claims of effectiveness. The therapy was often based on anecdotal accounts rather than rigorous scientific research.

**Safety Concerns:** The intensity of the emotional release in Primal Scream Therapy raised concerns about the potential for retraumatization or harm to individuals undergoing the treatment.

**Limited Application:** Primal Scream Therapy was seen as a highly specialized and niche approach that might not be suitable or effective for a wide range of mental health issues.

**Ethical Questions:** The therapeutic practice of reenacting traumatic experiences, especially in group settings, raised ethical questions about the potential for emotional manipulation or harm.

**The Decline of Primal Scream Therapy:** The decline of Primal Scream Therapy as a widely accepted therapeutic approach was gradual but significant. Several factors contributed to its diminishing popularity:

**Scientific Scrutiny:** As the field of psychology continued to evolve, there was a growing emphasis on evidence-based practices. Primal Scream Therapy, lacking empirical support, could not withstand scientific scrutiny.

**Alternative Therapies:** The emergence of more diverse and evidence-based therapeutic approaches, including cognitive-behavioral therapy, dialectical behavior therapy, and psychopharmacological interventions, provided viable alternatives for individuals seeking mental health treatment.

**Media Saturation:** While Primal Scream Therapy initially gained attention through media coverage and celebrity endorsements, it also faced backlash and parody, which contributed to its diminishing credibility.

**Regulation and Professional Standards:** The therapy's unconventional nature and concerns about potential harm led to increased scrutiny and regulation within the mental health profession.

**The Discrediting of Primal Scream Therapy:** Primal Scream Therapy gradually lost credibility as the mental health field moved towards evidence-based practices and away from unproven, emotionally intense therapies. By the late 20th century, it was no longer considered a mainstream or widely accepted approach to mental health treatment. Its decline can be seen as a reflection of the evolving standards and priorities within the field of psychology.

In conclusion, Primal Scream Therapy, a controversial and unconventional approach to mental health developed by Arthur Janov in the 1960s, gained attention and popularity but faced significant skepticism and criticism. While it attracted notable supporters, including celebrities, its lack of empirical evidence, safety concerns, limited applicability, and ethical questions led to its gradual decline. By the late 20th century, Primal Scream Therapy had lost credibility as a mainstream therapeutic approach, serving as a cautionary tale about the importance of evidence-based practices in the field of psychology.

# Malleability of Memory

The theory of the malleability of memory represents a significant milestone in the field of psychology, fundamentally altering our understanding of how memories are formed, stored, and retrieved. This theory challenges the notion that memories are stable and unchanging, suggesting instead that they can be modified, distorted, or even fabricated under certain conditions. In this exploration, we will delve into the origins of the malleability of memory theory, the researchers who pioneered this idea, the initial skepticism it faced, and the pivotal experiments that led to its widespread acceptance.

**Origins of the Theory:** The concept of the malleability of memory began to take shape in the mid-20th century, challenging the prevailing belief that human memory functions like a fixed and accurate recording of past events. Researchers started to explore how various factors, such as suggestion, social influence, and cognitive processes, could influence the formation and retrieval of memories.

**The Role of Elizabeth Loftus:** Dr. Elizabeth Loftus, a prominent American cognitive psychologist, played a pivotal role in advancing the theory of memory malleability. In the 1970s, Loftus conducted a series of groundbreaking experiments that demonstrated how suggestion and leading questions could distort eyewitness testimonies and shape individuals' recollections of events.

One of her most famous experiments involved showing participants a video of a car accident and subsequently asking them questions about what they had witnessed. By subtly altering the wording of the questions, Loftus was able to influence participants' memory recall. For instance, replacing the word "hit" with "smashed" led participants to recall the accident as more severe than it actually was.

**Prominent Believers and Challenges:** The theory of memory malleability faced skepticism and resistance, particularly from those who held onto the traditional view that memories were reliable and immutable records of past events. Many legal professionals, for example, were initially reluctant to accept the implications of memory research for eyewitness testimony in criminal cases.

However, as Elizabeth Loftus and other researchers continued to produce compelling evidence through controlled experiments, the idea that memory is malleable gained traction. Cognitive psychologists and neuroscientists increasingly recognized that memory processes were subject to various cognitive biases, errors, and influences.

**Experiments and Key Findings:** Several pivotal experiments contributed to the acceptance of the malleability of memory theory:

**Misinformation Effect:** Loftus's research on the misinformation effect demonstrated that exposure to misleading information after an event could distort an individual's memory of that event. Participants who were presented with false information were more likely to incorporate it into their recollection of the original event.

**Source Amnesia:** This phenomenon, also explored by Loftus, involves forgetting the source of a memory while retaining the content. People might remember an event but forget where or when they encountered the information, leading to confusion about the accuracy of the memory.

**Repressed and Recovered Memories:** Research in the 1980s and 1990s highlighted the potential for repressed and subsequently "recovered" memories of traumatic events. However, controversy surrounded the validity of these claims, and skepticism grew regarding the accuracy of such memories.

**Widespread Acceptance and Legal Implications:** The theory of memory malleability gradually gained widespread acceptance within the scientific community, prompting a paradigm shift in how psychologists and neuroscientists approached the study of memory. Researchers began to emphasize the reconstructive nature of memory, acknowledging its susceptibility to suggestion, leading questions, and cognitive biases.

This shift in understanding had significant implications for various fields, including law enforcement, legal proceedings, and therapy. Legal experts increasingly recognized the fallibility of eyewitness testimony and the potential for memory distortion in legal cases. Therapists and counselors also became more cautious about the retrieval and interpretation of repressed or traumatic memories, emphasizing the need for evidence-based practices.

**Discrediting the Theory and Continued Research:** It is important to note that the malleability of memory theory is not a theory that was "disproved" but rather one that evolved and matured through ongoing research. While the early experiments of Elizabeth Loftus and others highlighted the malleability and fallibility of memory, subsequent research has delved deeper into the underlying mechanisms of memory distortion and how to mitigate its effects.

Modern research in cognitive psychology and neuroscience continues to explore memory processes, including the role of the hippocampus in memory consolidation, the impact of emotional arousal on memory formation, and the factors that contribute to false memories. These investigations build on the foundational work of pioneers like Loftus and contribute to a more nuanced understanding of the complexities of memory.

**In Summary:** The theory of memory malleability, which challenged the traditional view of memory as a stable and accurate record of past events, originated in the mid-20th century. Dr. Elizabeth Loftus's groundbreaking experiments on the misinformation effect and source amnesia played a pivotal role in advancing this theory. Initially met with skepticism, the theory gradually gained acceptance within the scientific community, prompting a paradigm shift in the understanding of memory. While the theory itself was not disproved, ongoing research continues to refine and deepen our understanding of memory processes and their susceptibility to various cognitive biases and influences.

# Schizophrenogenic Mother

The concept of the "schizophrenogenic mother" is a discredited theory that once held a prominent place in the field of psychology, particularly in explaining the origins of schizophrenia. This theory postulated that the behavior and parenting style of mothers played a causative role in the development of schizophrenia in their children. In this exploration, we will delve into the origins of the schizophrenogenic mother theory, its proponents, the era when it gained prominence, the eventual debunking of the theory, and the impact of its discrediting on our understanding of schizophrenia.

**Origins of the Theory:** The schizophrenogenic mother theory emerged in the mid-20th century as an attempt to explain the origins of schizophrenia, a complex and debilitating psychiatric disorder characterized by disturbances in thought, emotion, and behavior. This theory was rooted in the broader field of psychoanalytic thought, which was influential at the time.

**Origins and Proponents:** While it is challenging to attribute the theory to a single individual, it was closely associated with the psychoanalytic school of thought, particularly the work of psychoanalyst Frieda Fromm-Reichmann. Fromm-Reichmann, a German psychiatrist, wrote extensively about schizophrenia and was a proponent of the idea that the family dynamics and maternal behavior played a significant role in the onset of the disorder.

**Believers in the Theory:** The schizophrenogenic mother theory gained traction within the psychoanalytic and psychiatric communities during the mid-20th century. Prominent figures in the field, including psychoanalysts and psychiatrists, endorsed and propagated this theory. Some believed that mothers who were perceived as controlling, cold, or domineering were contributing to the development of schizophrenia in their offspring.

**The Impact on Families:** The acceptance of the schizophrenogenic mother theory had profound and harmful consequences for families dealing with schizophrenia. Parents, especially mothers, were unfairly stigmatized and blamed for their children's mental health challenges. They faced criticism for their parenting style and were made to feel responsible for the onset of schizophrenia in their loved ones. This added emotional distress to an already difficult situation.

Families often experienced guilt, shame, and social isolation due to the prevailing belief that the behavior of mothers was a primary cause of schizophrenia. Moreover, this theory diverted attention away from the search for biological and genetic factors contributing to the disorder.

**Challenges to the Theory:** The schizophrenogenic mother theory faced increasing criticism as the field of psychology and psychiatry advanced, and empirical research became more central to understanding mental disorders. Several key factors contributed to the eventual debunking of the theory:

**Lack of Empirical Evidence:** The theory lacked robust empirical support. Research failed to consistently demonstrate a direct causal link between maternal behavior and the development of schizophrenia in offspring.

**Emergence of Biological and Genetic Factors:** As research into schizophrenia advanced, it became increasingly clear that biological, genetic, and neurobiological factors played a crucial role in the disorder. This shift in focus moved away from attributing schizophrenia solely to parenting style.

**Broader Understanding of Schizophrenia:** The view of schizophrenia evolved to recognize it as a complex and multifactorial disorder with a range of genetic, environmental, and neurodevelopmental influences. This perspective undermined the simplistic notion of maternal causation.

**Changing Paradigms in Psychiatry:** The field of psychiatry moved away from psychoanalytic explanations and embraced more biopsychosocial models of mental disorders, emphasizing the interplay of biological, psychological, and social factors.

**Discrediting the Theory:** The debunking of the schizophrenogenic mother theory was a gradual process that unfolded over several decades. By the latter half of the 20th century, the theory had lost its credibility within the scientific and psychiatric communities. Researchers and clinicians increasingly recognized the limitations of the theory and the need for a more nuanced and comprehensive understanding of schizophrenia.

**The Modern Understanding of Schizophrenia:** Contemporary research and clinical practice emphasize the multifactorial nature of schizophrenia. While genetic and neurobiological factors are recognized as critical, environmental and psychosocial factors also play a role in the development and course of the disorder. Current approaches to schizophrenia treatment and management are informed by a biopsychosocial model that considers the interplay of genetic vulnerability, neurodevelopmental factors, and environmental stressors.

**Conclusion:** The concept of the schizophrenogenic mother, once a prominent explanation for the origins of schizophrenia, has been discredited and relegated to the annals of history. This theory, rooted in psychoanalytic thought and the belief that maternal behavior was a primary cause of the disorder, had profound and harmful consequences for families affected by schizophrenia. The discrediting of this theory marked a significant shift in our understanding of schizophrenia, emphasizing the need for a more comprehensive and evidence-based approach that considers a range of genetic, neurobiological, and environmental factors in the development and course of the disorder.